ASSESSMENT AND PREDICTION OF SPEECH QUALITY IN TELECOMMUNICATIONS

Assessment and Prediction of Speech Quality in Telecommunications

by

Sebastian Möller
Institut für Kommunikationsakustik (IKA)

KLUWER ACADEMIC PUBLISHERS
BOSTON / DORDRECHT / LONDON

A C.I.P. Catalogue record for this book is available from the Library of Congress.

ISBN 0-7923-7894-6

Published by Kluwer Academic Publishers,
P.O. Box 17, 3300 AA Dordrecht, The Netherlands.

Sold and distributed in North, Central and South America
by Kluwer Academic Publishers,
101 Philip Drive, Norwell, MA 02061, U.S.A.

In all other countries, sold and distributed
by Kluwer Academic Publishers,
P.O. Box 322, 3300 AH Dordrecht, The Netherlands.

Printed on acid-free paper

Printed in the Netherlands.

DES MENSCHEN AUGE HAT NIE GEHÖRT,
DES MENSCHEN OHR NIE GESEHEN;
DES MENSCHEN HAND KANN NICHT SCHMECKEN;
SEINE ZUNGE NICHT BEGREIFEN
NOCH SEIN HERZ BERICHTEN,
WAS MEIN TRAUM WAR.

Ein Sommernachtstraum, IV, 1

Contents

Definitions and Abbreviations

Definitions

A	expectation factor
A_{LOI}	factor for calculating LOI, depending on the loudness of received speech
A_w	warping amplitude
a, b	constants of the 'power law'
B'	frequency weighting function related to loudness
B'_a	frequency weighting function related to articulation
B'_{LE}	frequency weighting function related to listening–effort
b_0	description of an auditory event
β_0	pure tone threshold of hearing in quiet [dB rel. 20 μPa·Hz$^{-1/2}$]
β'_S	spectrum density of speech at MRP [dB rel. 20 μPa·Hz$^{-1/2}$]
D	frequency–weighted version of $DELSM$
D_{LOI}	factor for calculating LOI, depending on the level of circuit noise
d	magnitude of a stimulus
d_M	magnitude of a stimulus at the midopinion value of the scale
$DELSM$	frequency–dependent difference in sensitivity between the directed and diffuse sound [dB]
%Diff	percentage of users experiencing difficulty in talking or listening over a connection
f, f_k	frequency [Hz]
F_s	sampling frequency [Hz]
F_w	warping frequency [Hz]
$(\Delta f)_c$	bandwidth of critical band [Hz]
G	signal–to–equivalent–continuous–circuit–noise ratio [dB]
g	exponent of the logistic psychometric function
G_L	frequency weighting function for the calculation of loudness ratings
%GoB	percentage of users rating a connection good or better
h_0	auditory event
$I, Itot$	impairment factor
Id	impairment factor for delayed impairments related to the speech signal
Ie	equipment impairment factor
Iq	impairment factor for quantizing distortion
Is	impairment factor for impairments occurring simultaneously with the speech signal

I_{QoS}	GSM quality of service index
K	factor for calculating Y_C from Y_{LE}
k	exponent of the relation between subjective rating and apparent magnitude
K_C	allowance to the threshold of hearing of complex tones in quiet [dB]
K_i	frequency weighting function for alternatively calculating loudness ratings
l_k	point of maximum excitation on the basilar membrane measured from the helicotrema due to a tone at f_k [mm]
l_{max}	total length of the basilar membrane [mm]
Le	frequency–dependent loss of the talker echo path [dB]
Lst	frequency–dependent loss of the sidetone path [dB]
L_{ME}, L_{RME}, L_{UME}	air–to–air transmission loss from mouth reference point to ear reference point [dB]
$\overline{L_{RME}}$, $\overline{L_{UME}}$	weighted average mouth–to–ear loss [dB]
λ, λ_R, λ_U	impression of loudness
LOI	listening opinion index
$LSTR$	listener sidetone rating [dB]
L_0	constant factor for calculating loudness ratings
$(\Delta l)_c$	critical length [mm]
M	shift in hearing threshold attributable to the presence of noise [dB]
m	exponent of the growth function $Q(Z)$
mse	mean squared error
$n(k)$	sampled noise signal
N	total number of frequency bands for calculating loudness ratings
Nc	circuit noise level [dBm0p]
$Nfor$	noise floor level [dBmp]
N_P	perceptual magnitude
No	total equivalent circuit noise level [dBm0p]
OLR	overall loudness rating between MRP and ERP [dB]
p, p_r, p_s	mean normalized opinion for an impairment
$P(Z)$	growth function of Z related to listening–effort
$P_a(Z)$	growth function of Z related to articulation
Pr	A–weighted sound pressure level of room noise at receive side [dB(A)]
Ps	A–weighted sound pressure level of room noise at send side [dB(A)]
%PoW	percentage of users rating a connection poor or worse
Q	signal–to–quantizing–noise ratio [dB]

$Q(Z)$	growth function of Z related to loudness
qdu	quantizing distortion unit
r	Pearson correlation coefficient
R	transmission rating
R_{CC}	transmission rating taking into account call–completion impairments
R_{CS}	transmission rating taking into account call–setup impairments
R_{LNE}	transmission rating taking into account loss, noise and talker echo
R'_w	weighted sound reduction index [dB]
Ro	basic signal–to–noise transmission rating factor
RLR	receive loudness rating between the 0 dBr point in the network and the ERP [dB]
$RLRset$	receive loudness rating of the telephone handset [dB]
s_0	sound event
S_{JE}	electro–acoustic receiving sensitivity from junction to ERP [dB]
S_{MJ}	acousto–electric sending sensitivity from MRP to junction [dB]
S_{meST}	air–to–air sensitivity of the electric sidetone path for directed (speech) sound, as defined in ITU–T Rec. P.64 (1997) [dB]
S_{RNST}	air–to–air sensitivity of the electric sidetone path for diffuse (room noise) sound, as defined in ITU–T Rec. P.64 (1997) [dB]
SLR	send loudness rating between the MRP and the 0 dBr point in the network [dB]
$SLRset$	send loudness rating of the telephone handset [dB]
$STMR$	sidetone masking rating [dB]
t	level of sensation on a continuous finite rating scale
T	mean one–way talker echo path delay [ms]
Ta	overall delay between MRP of the talker and ERP of the listener [ms]
Tr	round–trip delay for listener echo [ms]
$TELR$	talker echo loudness rating [dB]
%TME	percentage of users terminating a call early
V_C, V_L	active speech level in conversation or listening–only situations [dBV]
W_i	frequency weighting function for calculating loudness ratings
$WEPL$	weighted echo path loss for listener echo [dB]
$x(k)$	sampled input signal

$y(k)$	sampled output signal
Y_{LE}	listening–effort mean opinion score
Y_{LP}	loudness–preference mean opinion score
Y_{LQ}	listening–quality mean opinion score
Z, Z_R, Z_U	sensation level [dB]
Z_L	sensation level in the absence of noise [dB]
Z_{R0}	sensation level Z at a given frequency when $L_{ME} = 0$ [dB]

Abbreviations

ACELP	algebraic code–excited linear prediction
ACR	absolute category rating
ADPCM	adaptive differential pulse code modulation
AEN	articulation reference equivalent
AoS	acceptability of (a) service
ATM	asynchronous transfer mode
BcTR	Bellcore transmission rating model
BP	bandpass
CA	composite acceptability measure
CCI	call clarity index
CCITT	Comité Consultatif International Télégraphique et Téléphonique
CCR	comparison category rating
CLID	cluster identification test
CS–ACELP	conjugate structure algebraic code–excited linear prediction
CT	conversation test
CVC	consonant–vowel–consonant cluster
[dBm]	logarithmic measure of the magnitude of a signal
[dBm0]	logarithmic measure of the magnitude of a signal, referred to the 0 dBr reference point of the network
[dBmp]	logarithmic measure in [dBm], but weighted with a psophometric weighting as described in ITU–T Rec. O.41 (1994)
[dBm0p]	logarithmic measure in [dBm0] referred to the 0 dBr reference point of the network, but weighted with a psophometric weighting as described in ITU–T Rec. O.41 (1994)
[dBr]	logarithmic measure of signal magnitude relative to a virtual point designated a level of 0 dBr in the network
[dBV]	logarithmic measure relative to 1 V
DAM	diagnostic acceptability measure
DAT	digital audio tape

DC	direct current
DCME	digital circuit multiplexing equipment
DCR	degradation category rating
DECT	digital European cordless telecommunication
DRT	diagnostic rhyme test
DSP	digital signal processor
EC	echo canceller
ERP	ear reference point as defined in ITU–T Rec. P.64 (1997)
ETSI	European Telecommunications Standards Institute
FIFO	first–in first–out
GLS	Graeco–Latin square
GSM	global system for mobile communications
GSM–FR	GSM full rate codec
GSM–HR	GSM half rate codec
GSM–EFR	GSM enhanced full rate codec
HATS	head and torso simulator
HFT	hands–free terminal
HP	highpass
IKA	Institute of Communication Acoustics
INDSCAL	individual differences scaling
INMD	in–service non–intrusive measurement device
IP	internet protocol
IRS	intermediate reference system as defined in ITU–T Rec. P.48 (1989)
ISDN	integrated services digital network
ISO	International Organization for Standardization
ITU	International Telecommunication Union
ITU–T	International Telecommunication Union — Telecommunication Standardization Sector
LD–CELP	low–delay code–excited linear prediction
LOT	listening–only test
LP	lowpass
MCS	mobile communication system
ME	magnitude estimation
MELP	mixed excitation linear prediction
MNRU	modulated noise reference unit as defined in ITU–T Rec. P.810 (1996)
MOS	mean opinion score (often used synonymously to Y_{LQ} or Y_C)
MRP	mouth reference point as defined in ITU–T Rec. P.64 (1997)
MRT	modified rhyme test
OPINE	overall performance index model for network evaluation
PBS	packet based system

PC	paired comparison technique
PCM	pulse code modulation
PR	paired rating technique
PSQM	perceptual speech quality measure according to ITU–T Rec. P.861 (1996)
PSTN	public switched telephone network
QoS	quality of (a) service
rms	root mean square
RPE–LTP	regular pulse excitation long term prediction
SCT	short conversation test
SNR	signal–to–noise ratio
SRAEN	système de référence pour la détermination de l'affaiblissement équivalent pour la netteté
SSRCT	single stimulus rating conversation test
SUBMOD	subjective model
TFW	time and frequency warping
TIPHON	telecommunication and internet protocol harmonization over networks
TOSQA	Telekom objective speech quality assessment
TTS	text–to–speech
UPT	universal personal telecommunication
VAD	voice activity detection
VoIP	voice over internet protocol
VSELP	vector sum excited linear prediction

Acknowledgements

The present work was performed during my occupation at the Institute of Communication Acoustics, Ruhr–University Bochum. It would not have been possible without the help of many supporters. Especially, I would like to thank the following:

- my colleague Dr. Ute Jekosch for many fruitful discussions, help and support over the years,
- the head of the institute, Prof. Dr.–Ing. Dr. techn. h.c. Jens Blauert for enabling this work and supporting it with interest and advice,
- Prof. Dr.–Ing. Ulrich Heute for kindly taking over the co–supervision, and for his interest in my work,
- my long–term student co–worker Joachim Riedel for his work on the test organization and his comprising support,
- Susanne Krause and Sabine Jekosch for the organization and run of numerous auditory experiments,
- many other colleagues of the Institute of Communication Acoustics for providing a very good working atmosphere, and for a fruitful and friendly collaboration,
- Carol Hill, Dr. Ute Jekosch and Dr.–Ing. Klaus Rateitschek for reviewing the manuscript,
- Stephan Wiegelmann, Ralf Plücker, Joachim Riedel and Paul Darmoe for their work on diploma thesis and study projects related to this work,
- the system administrators of the institute's computer network and the members of our office, for providing and maintaining their resources,
- Dr.-Ing Jens Berger and other colleagues within the ITU–T Study Group 12 for their interest and critics in my work, and
- my family and many friends for supporting me over the last years.

Prof. Gunnar Borg kindly agreed the reproduction of his CR–10 scale which is copyrighted by him. Figure 3.5 is an extract of ITU–T Rec. G.107 (1998) and has been reproduced with the kind authorization of the ITU who is the copyright holder. The sole responsibility for selecting extracts for reproduction lies with the author alone and can in no way be attributed to the ITU. The implementation of the SUBMOD model was kindly provided by British Telecom. The citation is a reproduction of Ralf Königs "Jago".

Summary

The quality of a telecommunication voice service is largely influenced by the quality of the transmission system. Nevertheless, the analysis, synthesis and prediction of quality should take into account its multidimensional aspects. Quality can be regarded as a point where the perceived characteristics and the desired or expected ones meet. A schematic is presented which classifies different entities which contribute to the quality of a service, taking into account conversational, user as well as service related contributions.

Starting from this concept, perceptively relevant constituents of speech communication quality are identified. The perceptive factors result from elements of the transmission configuration. A simulation model is developed and implemented which allows the most relevant parameters of traditional transmission configurations to be manipulated, in real time and for the conversation situation. Inputs into the simulation are instrumentally measurable quality elements commonly used in transmission planning of telephone networks. A reduced set of these quality elements forms a basis for models which aim at predicting mouth–to–ear quality as it would be perceived by a user of the system. These models are an important tool for the planner of telecommunication networks, as they allow the expected quality to be estimated in advance, even before the network has been set up. Two well–known models (the SUBMOD and the E–model) are analyzed in more detail, with an emphasis on the psychoacoustic and psychophysical backgrounds.

Valid measurements of speech communication quality and usability can only be obtained by performing tests with human subjects. More precisely, the assessment of speech communication quality requires a conversation situation. In order to avoid the drawbacks of traditional conversation tests, a more economical method called 'short conversation test' (SCT) is developed. Tests carried out using SCT scenarios are a powerful tool for assessing integral speech communication quality, provided that appropriate scaling methods are available. Considering some major disadvantages of the traditional 5–point quality scale (MOS scale), a proposal is made for scaling on an impairment scale, which shows several similarities to the scale which forms the basis of the E–model. For the first time, the method allows E–model 'impairment factors' for new types of equipment to be measured in a direct way.

Using the real–time simulation of the telephone transmission implemented in this work, a large number of auditory experiments is carried out in order to investigate the interdependence of factors governing the quality of a service. Results show that the user background (expectation, references) has a strong influence on quality that is interpreted at an 'absolute' level. The quality of prediction models is further investigated by comparing instrumentally predicted quality to the results of the performed auditory tests. It turns out that model predictions are satisfactory for many types of degradations. For other impairments, the models can still be improved by modifying the model algorithm, or with the help of other instrumental models (e.g. for codecs). Proposals are made for this in order to achieve more valid quality predictions for modern and future telecommunication scenarios.

Chapter 1

INTRODUCTION

"So near ... as if you were there!" "Excellent speech quality in the high–quality net." "You can now make phone calls where the quality is as good as in usual wireline networks." "The GSM 1800 standard provides speech transmission with exceptional clarity and brilliance." These extracts from German telecommunication provider advertisements are some examples showing that quality has become an important marketing factor in telephony. Apart from the terminal equipment and service novelties, it is the speech communication quality which has regained a considerable importance. This is astonishing because the latter has been taken for granted for many years, as long as intelligibility was no longer a problem in ordinary wireline telecommunication networks. Though new technologies have been introduced and a steady transition is made towards fully digital circuits, it seems that speech communication quality tended to decrease again in the mid–term.

One major reason for this decrease is the advent of new technologies, which made it necessary to make more economical use of the limited transmission channels. This is especially the case for mobile communication systems, where it is generally accepted that such systems provide a distinctly lower level of speech transmission quality. Draft ETSI Guide EG/STQ-00004 (1998) gives three main sources why speech communication quality has deteriorated within the last years: the evolving technology introduces new types of impairment, whereas traditional faults are getting less significant; the more complex network technology; and less control over what happens in modern public/private networks due to the liberalization of the telecommunication market. Nevertheless, high sales prove that mobile phones enjoy a wide acceptance in many industrial countries.

It has become obvious that the quality of a telecommunication service is not only determined by the (physical) factors which may exercise an influence on the speech transmission itself. Instead, the user of a service makes an indirect evaluation or trade–off between what he/she expects for that specific situation of telephone use, and what is provided by the service. Bearing that in mind, several questions arise: What is speech quality in the context of a telecommunication service? Does a specific service (e.g. a mobile service) appear to have a higher quality for its users though offering an (auditorily) more severely degraded transmission? Which factors that influence the quality of a service are known, and how are they related to each other? Can compromises be made between such factors, resulting in an optimum quality for a specific situation? For which group of users will such a compromise be valid?

The word 'telecommunication' originates from the Greek word *tele* — meaning from afar — and the Latin *communicare*, to share. Telecommunication thus serves to share, transmit or exchange information between two (or more) remote parties. A telecommunication situation is most often a special class of a human–human interaction, in which both humans are located at a man–machine interface. Although located at such an interface, it is generally clear to the user that he/she is conversing with a human partner on the other end of the line, and not with a machine. The human behavior reflects this fact, and so does the human expectation of the service. Because he/she has experience of operating such a system, the user of a telecommunication service will have a certain idea of what quality can be expected beforehand. The situation is different for new types of services, such as automated dialog systems for information retrieval, announcements using synthesized speech, etc., where experience has not been stabilized yet. The quality of a standard telecommunication service for voice transmission is linked to the communication situation. Quality, as defined in the following chapter, is not something that results directly from the physical characteristics of the transmission line; rather, it is attributed to the system in a specific situation by the user.

As a consequence, quality assessment of telecommunication services has to reflect a realistic communication situation. In the first instance, the components or dimensions of quality have to be identified for a specific service. Then, methods and methodologies have to be developed which permit reasonable estimates of such dimensions to be obtained. The normal usage of a telephone connection serves to exchange information in both directions, so the conversation situation is the only realistic assessment situation of such services. If preference is given to listening–only tests (e.g. because of time or money constraints), the lack of a natural situation may exercise an influence on the quality assessment results, and has to be taken into account when interpreting test results.

Generally speaking, two different approaches exist to determine speech quality. The more general and the only valid one is to obtain auditory judgments — in listening–only or conversation tests, user surveys, etc.— from human subjects. Because it is the user who establishes a relationship between the expected or desired and the characteristics offered by the service, there is generally no way of performing reasonable quality assessment without relying on test subjects. Questions arising in auditory speech quality evaluation include: Which dimensions exercise an influence on the quality of a service? Which methods or methodologies exist to determine such dimensions? Is it possible to run laboratory tests under controlled conditions, or is it necessary to perform actual users' surveys? What type of information can be obtained from such experiments?

Apart from auditory testing, instrumental methods or models exist which aim to predict overall quality or specific quality dimensions based on physically measurable parameters or signals. These models produce reasonable estimates for a limited set of parameters and for a limited test condition, namely the one which has been used to establish the model. They are especially important for the planner of telecommunication networks who wishes to have reasonable estimates of the expected quality even before a network has been set up. A standard problem in engineering arises: a trade–off between optimum network quality and lowest possible costs has to be made, in order to avoid acceptability problems (due to low speech transmission quality) as well as over–engineering. Models which provide reasonable predictions of future speech communication quality are a valuable tool for the network planner enabling him/her to find such a trade–off for the specific system. Many of them have been developed over the past years and include a high amount of experience with existing networks. In the last decade, strengths have been concentrated in order to define one single network planning model which can provide quality estimates for mixed, analog–digital networks, including a couple of new types of equipment encountered in such networks, e.g. low–bitrate codecs. This model is now called the E–model and has been standardized by the International Telecommunication Union (ITU) and the European Telecommunications Standards Institute (ETSI). Other approaches exist which make use of frequency–dependent input information, such as the SUBMOD model developed by British Telecom. Although these models have been in use for a number of years and seem to provide reasonably accurate quality predictions, they have never been fully verified. In particular, it is not clear what dimensions of quality such models' predictions refer to. The validity and precision of quality predictions, depending on the input parameter constellation, also have to be analyzed.

In the past, quality levels for ordinary telephone connections were termed "toll quality" for the public switched telephone network (PSTN) or "speech communications quality" for mobile systems. These terms were used without a precise definition of the affiliation to one particular class. Recently, new approaches have been launched to define classes of "speech transmission quality" based on quality estimates of network planning models (ITU–T Draft Rec. G.109, 1998), and further related to expected "user satisfaction". Because many network planners will make use of this classification, it is important to know which aspects of quality are catered for by the respective quality prediction models. A more general concept of quality will uncover further dimensions which play an important role in forming a quality percept, but which are not yet included in modeling approaches.

It is the aim of this work to examine the questions raised above. The approach made here is to define a new analytic concept of quality of a service, which takes into account other, more user–directed concepts, such as usability, utility and acceptability (Chapter 2). Making use of that concept, several constituent dimensions of the quality of a service are identified. The quality dimensions are related to both perceptive and physical factors describing the auditory conversation event and the characteristics of the telephone connection. In order to render the effects of physical factors perceivable, a real–time telephone connection simulation model is developed and implemented, which enables carrying out conversations with defined settings of the transmission circuit mouth–to–ear. This is done in more detail in Chapter 3. On the other hand, assessment methods and methodologies can be identified which allow different quality dimensions to be assessed, see Chapter 4. The assessment methods are auditory in principle, because the definition of quality in the concept established here requires the use of test subjects for building the relation between expected and offered characteristics. New test scenarios are developed and appropriate scaling methods are discussed, enabling to obtain highly informative quality assessment results in natural conversation situations. The models for predicting speech communication quality based on instrumentally measurable characteristics of the connection are discussed in more detail in Chapter 5. The schematic established at the beginning allows all — perceptive factors, factors relating to the telephone connection, assessment methods and quality prediction models — to be classified according to the quality features which will be experienced by the user. This classification is very helpful as it gives information on which quality features are affected by a specific characteristic of the transmission, and how the effect can be auditorily measured or even predicted. In Chapter 6 relations between different factors which govern the quality schematic are discussed. The bases of discussion are well–selected auditory test results, carried out as a part of the present work and using the new telephone line simulation established here. Auditory tests also permit quality estimates obtained from prediction models to be analyzed. A comparison is made in Chapter 7, with an emphasis on the E–model which is a distillate of several precursors and the only one recommended by the ITU–T for use in network planning. A final discussion and an outlook regarding the future development of quality prediction models for telecommunication scenarios conclude the considerations in Chapter 8.

It is hoped that this book will meet the particular interests of engineers working in quality management and planning of voice transmission networks, within standardization bodies (ITU–T, ETSI) and elsewhere. A broad basis in psychoacoustics as well as in quality assessment seems to be necessary for quality planning and quality prediction, a fact which makes Chapter 4 somewhat long. On the other hand, the necessary fundamentals in telephony are discussed in Chapter 3. It is hoped that the more general approach used in this work will help to make careful and well–understood use of quality assessment methods. The second part (Chapters 5 to 7) may contribute to a further clarification of the aims and limitations of network planning models producing quality estimates mouth–to–ear.

Chapter 2

CONSIDERATIONS ON QUALITY

1. SPEECH QUALITY

Language is the primary system of signs, used as a tool for thinking, acting, and communication (Lewandowski, 1980). Speech, as opposed to language, is the application of this system of verbal signs, in order to express and transmit information. In the sense of Peirce (1998) and Ogden and Richards (1960), a sign is something which represents something else and is understood by someone, thus a triadic relationship between (1) a verbal, visual or acoustic form, (2) a signed object (content) and (3) the interpreter (Lewandowski, 1980). When the quality of speech is investigated, is is very useful to differentiate between the form or surface structure, i.e. the acoustic presentation, and the contents or meaning. In the context of this work speech quality refers to the acoustic form of speech, in the sense that the acoustic presentation is the object of interest. However, the content has a strong influence on the perception of speech, and consequently on how a certain quality is attributed to the acoustic form. No natural relationship exists between the form (surface structure) and the contents. It is the human talker and listener who establishes this relationship.

In the context of speech perception, it is helpful to differentiate between a sound event and an auditory event (Blauert, 1997). When speech is perceived by a human being, the acoustic speech signal presents a sound event (s_0). The sound event is perceived by the human auditory system at different processing levels. The perception process results in an auditory event (h_0) which takes place in the auditory space. Unfortunately, the auditory event is private and there is no way of directly assessing it. This can only be done via the description level. Using introspection, the human being establishes a relationship between the auditory event and the description b_0. It is this process which takes place in a test situation when a test subject is asked to give his/her opinion on a stimulus. The situation can be represented as proposed by Blauert (1997), see Figure 2.1. The difficulty in investigating auditory percepts is a consequence

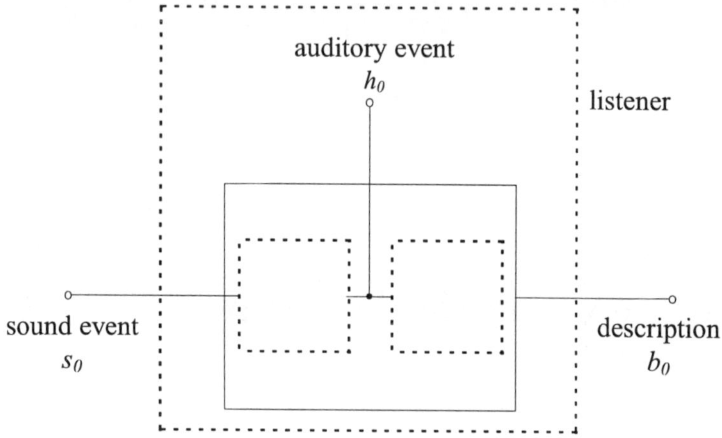

Figure 2.1. Schematic representation of the listener in an auditory experiment (Blauert, 1997).

of the impossibility of directly assessing the object of interest. Speech signals initiate the speech perception process, and even the resulting speech percepts can only be investigated via the description level. The form of the description is important for the type and amount of information one can obtain from test subjects as the result of an assessment process (Jekosch, 1998). The perception process is an instantaneous and individual event. There is a multitude of influencing factors within the perceiving person and in his/her environment (experience, expectation, emotions, motivation, etc.) which render the process unreproducible. Therefore, the term 'event' is preferred in this context: in a first place for the 'sound event' and the 'auditory event', and later (in accordance with Jekosch, 1998) for the 'quality event'.

When perceiving, the human being establishes a relationship between the form and the content of speech. Obviously, the listener who acts as an interpreter is also an active part of this relationship. Using the sign theory cited above, the perception of speech and of sound in general can be displayed by using the semiotic triangle, cf. Dürrer and Jekosch (1998). The general form of the semiotic triangle is reproduced in Figure 2.2. With regard to the content a speech signal is to carry, there are conventions and rules which are applied in the speech perception process. With regard to the form, however, there is only individual experience (borderlines) where the form is not appropriate to carry the meaning the speaker intends to convey. This differentiation between form and content is extremely helpful in the context of speech quality assessment. The assessment should not exclusively refer to the relation between form and interpreter, but has to take into account that the content exercises an influence on the speech percept as well. Assessment methods which neglect the content aspect are limited in their validity. Proposals for methods taking the content into account are further discussed in Section 9 of Chapter 4. In doing so, the whole situation in which the auditory event happens is interpreted. Speech quality is the judgment result of the entire speech percept, i.e. form and content.

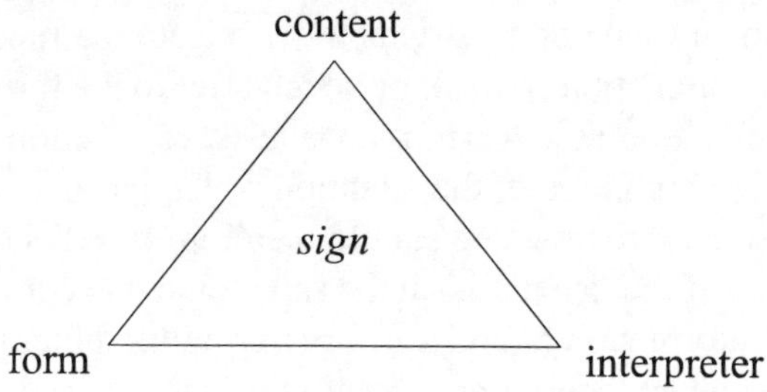

Figure 2.2. Semiotic triangle.

Jekosch (1998, p. 14) gives the following definition of quality, which will be used throughout this work:

> "Quality is the result of the judgment of a perceived constitution of an entity with regard to its desired constitution. [...] The perceived constitution contains the totality of the features of an entity. For the perceiving person it is a characteristic of the identity of the entity."

Applying this definition to speech, speech quality can be regarded as the result of a perception and assessment process, during which the assessing subject establishes a relationship between the perceived (i.e. the auditory event) and the desired or expected. Thus, the quality of speech does not exist in an absolute sense. It is attributed by the recipient, as the result of an assessment process (Jekosch, 1998). The term 'quality event' seems to be appropriate for describing this situation.

In a general dialog situation, the listener does not distinguish between surface and content. There are situations when the attention of the listener is more directed towards the contents (e.g. when thinking about what has been said) or towards the form. The latter may occur when the form of the speech signal is so bad that intelligibility suffers from it. In telephony, the form might be affected by e.g.

- a 'bad' connection,
- ambient (room) noise,
- the speaking style, e.g. a foreign accent,
- a listening disability, or

- a bad handset position.

Within the focus of the present work the quality of communication systems is to be assessed by means of transmitted speech signals. Because of this, speech quality refers to the acoustic realization of speech, and not to the content (what is said). Assessment methods exist which investigate the functional aspect of speech (e.g. task completion scores), or which refer to the form only indirectly by means of the function (e.g. performance tests, cf. Section 4 of Chapter 4). Though the form is highlighted, the attention of the listener will play a major role in the assessment process, and should therefore be reflected by the choice of the test method. If the listener is, at the same time, the conversation partner, the exchange of information happens in a bi– or multi–dimensional way. Both listener and talker adapt their communication behavior in order to optimize the communication efficiency. Examples of this behavior include

- question and reconfirmation (content),
- an adaptation of the speaking style (Lombard effect, communicating to foreigners or hearing–impaired), or
- an adaptation of the handset position.

Principally, speech quality can be approached from two sides. Until now, the quality and its perceived features have been regarded from the user's point of view. If the interest is on a communication system or service, or on a product in general, quality and its forming elements can also be regarded from the system designer's point of view. This 'planning quality' is, of course, related to the quality perceived by the communication partner. Both share the properties of the communication system. However, the quality features cannot be regarded as one–to–one transformations of the planner's quality elements. Such a cause–and–effect relation can, if it is ever possible, only be established for a specific product, system or service, and only with the help of the assessing subject or communication partner.

For the planning process of product quality, it is important to have an idea of the quality that would be perceived by the user in advance. Therefore, the prediction of quality becomes increasingly important. The prediction models, which are discussed in more detail in Chapter 5, can be regarded as a principal tool for the planner of speech communication systems to perform such predictions. One danger exists, namely that quality estimates of such models are regarded as more–or–less 'absolute' estimates of quality for an average user, or even as 'objective' estimates. This cannot be performed by prediction models, and was never intended by their developers. It is true that the large amount of different auditory test results leads to a certain degree of inter–subjectivity of the models' quality estimates. On the other hand, quality events can only be assessed in relation to the communicating subject's needs. Therefore, it is important to have specifications on the user's motivations, emotions and attitudes. The auditory tests performed in the context of this work in order to investigate the effect of user expectation (Chapter 4.) are a good illustration for this necessity. Needs are individual, but not necessarily contradictory, and for several cases they can be classified in order to reach a certain degree of homogeneity. This is important for system development, because only in very rare cases the planner of a communication system can satisfy specific individual user's requirements. In most cases, the planner of communication systems or services will try to satisfy the needs of a large number of potential users. It is the task of user surveys to investigate and describe these needs in as much detail as possible, in order to be able to plan a communication system in accordance with the users' needs. Only in this way can high quality communication systems be made.

2. QUALITY IN THE CONTEXT OF TELECOMMUNICATIONS

The basic considerations on speech quality can now be applied to telecommunication scenarios. A telecommunication system is generally constituted by a user terminal and the network and switching equipment. A service is defined by the use of such a system to fulfill a specific task (Gleiss, 1992).

There are different terms which denote aspects of quality as used in telephony. 'End–to–end quality' or 'mouth–to–ear quality' (German: *'Ende–zu–Ende–Qualität'* or *'Über–alles–Qualität'*) describe a quality which is related to the whole communication system, from the mouth of the talker to the ear of the listener. The first impression which the user of such a system has is called 'apparent quality' (German: *'Anmutungsqualität'*). Another term refers to a concept of quality which is constituted from different components or dimensions. The 'integral quality' (German: *'Gesamtqualität'*) describes a quality which reflects all such dimensions. Very often in telephony the term 'overall quality' is used, mainly as an equivalent to 'integral quality', but sometimes synonymously with 'end–to–end quality'. To avoid ambiguities the term 'integral quality' is preferred here.

ITU–T Rec. E.800 (1994) gives a definition of quality of service which is widely used in the telecommunication framework:

> "Quality of Service (QoS): The collective effect of service performance which determines the degree of satisfaction of a user of the service."

This definition is stated more precisely by defining four factors which characterize QoS:

- Service support: indicates how well a service provider or network operator is able to offer a service and to assist in its utilization. This performance depends mainly on the service provider organization.

- Service operabilityservice,operability: indicates how well a service is adapted to a successful and easy application by the user. It depends mostly on the network, terminal equipment, etc.

- Serveability: indicates to what extent a user can obtain a service and retain it for a desired period, without excessive impairment. Serveability includes service accessibility (the ability of a service to be obtained), service retainability (the probability that a service, once obtained, will continue for a given time duration) and service integrity. Service integrity indicates the degree to which a service is provided without excessive impairments. Serveability depends on the network, especially on how well the network can handle a certain amount of traffic, and which impairments are introduced by the network.

- Service security: indicates the protection provided against unauthorized monitoring, fraudulent use, malicious impairment, misuse, human error and natural disaster.

These quality of service aspects can be related to network performance concepts, namely trafficability and charging performance. The quality of service concept is useful for aiding a network operator or service provider to define, measure and control network characteristics (in terms of quality elements) which may help him/her to achieve a satisfactory level of service quality. However, it does not provide deep insight in the question of which quality features result in an acceptability of the service from the user's viewpoint. It is therefore necessary to install a quality of service schematic which is more user directed. This schematic was developed by the author and is illustrated in Figure 2.3. It will be presented and discussed in the following.

The quality of service concept used here refers to various types of telecommunication services, e.g. wireline telephony, wireless or mobile telephones, a speech communication service operating from an internet terminal, etc. Services which are not within its scope are e.g. responding machines or dialog systems for information retrieval. The quality of service is thought to be divided into three parts. The major component is called speech communication quality. It refers to the quality concept given above, and relates to a bi– (multi–) directional conversation over the telecommunication system. A second component is the service–related influences, which will be called the 'service performance' here. A service is generally established by a service provider, which — in the past — was often identical to the network operator. The 'service performance'

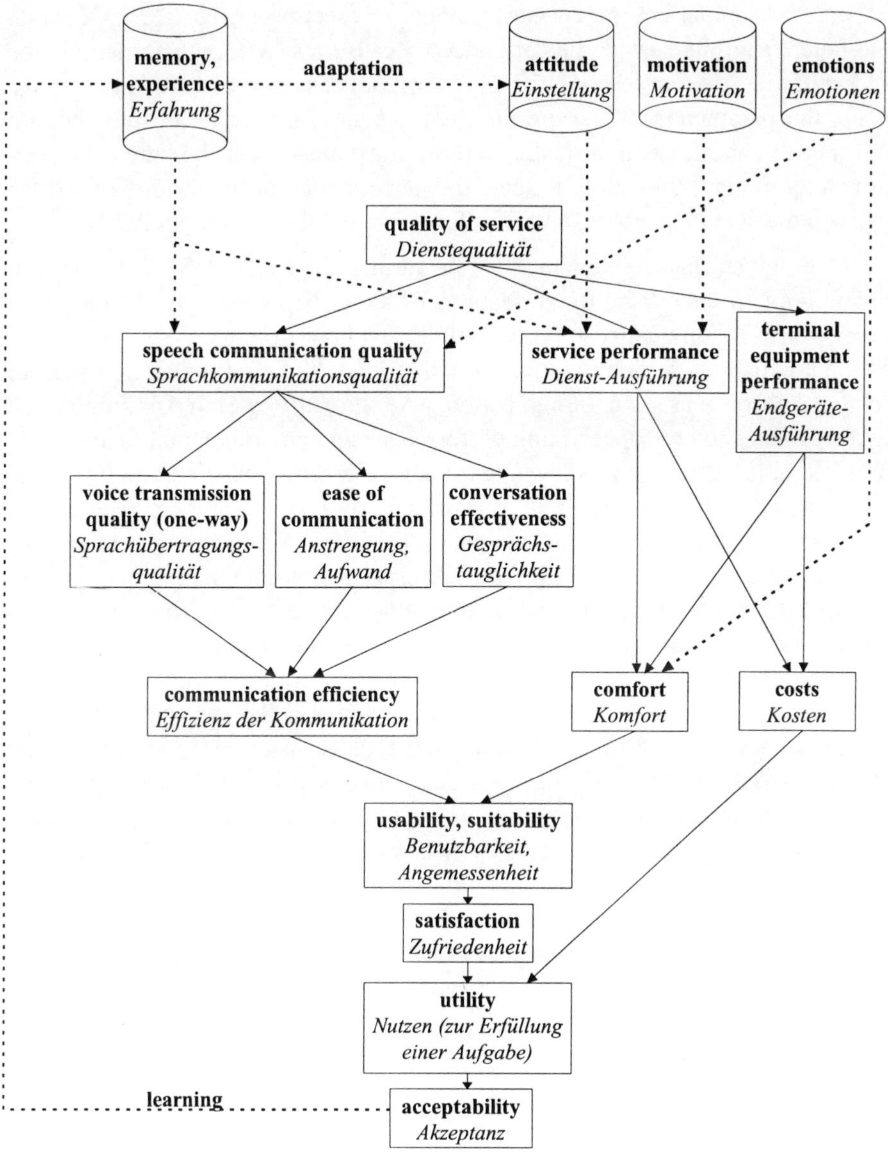

Figure 2.3. Quality entities describing the quality of service (QoS) and influencing the acceptabiliy of telecommunication speech services (user directed). German terms are given in italics.

component includes service support, a part of service operability and serveability, and service security as defined above. A third component of the QoS is the necessary terminal equipment performance. It seems appropriate to separate the latter from the service performance, because a service can sometimes be accessed via different terminals, or one terminal can operate different services. Also, the parameters of the terminal equipment which have an influence on the speech communication quality will be subsumed under the latter. The terminal equipment, thus, only reflects the performance of the terminal used for operating a specific service, and thus forms a part of service operability.

The given quality elements can be further subdivided. Speech communication quality is related to the features of the auditory event itself (one–way voice transmission quality, quality features resulting from the system's conversational capabilities (conversation effectiveness), and communication partner related features (ease of communication). Additionally, speech communication quality relates to both speech link performance and environmental factors (e.g. ambient noise, etc.). At this level, the quality elements (which are in the hands of the quality planner) are related to the quality features as perceived by the user. In the lower half of the schematic, the user–related features of quality are presented in relation to each other. All speech communication quality constituents (one–way voice transmission quality, conversation effectiveness and ease of communication) contribute to communication efficiency. Subsequently, communication efficiency related to the comfort experienced with a specific service together form the construct of usability, which is the appropriateness of a system or service to fulfill a defined task. Finally, usability in relation to the (financial) costs results in the utility of a system or service. The user may accept or refuse the utility which he/she experiences when using the telecommunication service. How readily a user uses the service is called the acceptability of a service. It can be measured by counting the users of a service and comparing this number to the number of potential users. Thus, acceptability is a purely economic concept. This is reflected in the new definition of the acceptability of a service given in EURESCOM Project P.807 Deliverable 1 (1998):

> "Acceptability of Service (AoS): Multidimensional property of a service, describing how readily a customer will use the service, represented as the ratio of the number of potential users to the quantity of the target group."

Both 'quality' and 'usability' or 'utility' are relative concepts, as they represent the meeting or trade–off between performance and the users' objectives. Usability can also be defined so that it presents the performance of the system or service in relation to the user's attitude (Gleiss, 1992). Speech quality can be defined as the result of the user's judgment on spoken language which he/she perceives in a specific situation and judges instantaneously according to his/her experience, motivation and expectation (Jekosch, 1998). Both quality and usability judgments are restricted to a specified user, task and environment. Therefore, the assessment of these concepts also has to conform to such restrictions. Only the inter–individual properties of assessment results can be translated from the situation they have been obtained in into other, more general situations. For practical applications, obviously, these inter–individual properties are of major interest, because they are a prerequisite for planning according to an unknown user's needs.

The schematic presented here differs in several points from the usability concept presented by Gleiss (1992). The main difference is that here QoS is defined in a user–directed way. The defining entities were chosen in a way that they can later be linked to assessment methods, physical characteristics and prediction models. As the term 'usability' refers to a similar relative concept as 'quality' does, it is symmetrically opposed to quality in the lower half of the schematic. The concept of 'attitude' in the usability definition of Gleiss (1992) is a rather diffuse one, and refers more to the assessment methodology than to a usability dimension. Here, the word 'comfort' is used instead. 'Comfort' means the ergonomic design of the terminal as well as the availability, reliability, call set–up time, etc. It also includes social presence ("warm, personal, sensitive, sociable") and aesthetic appeal ("colorful, large, spacious, beautiful, interesting"), resulting from emotions and attitude (Gleiss, 1992). The concept of acceptability, in contrast to Gleiss' definition, is regarded as a purely economic one. The term 'satisfaction' characterizes the user's state of mind. In the above schematic, it is understood as the result of the usability of the service, but it could easily be related to other quality features. Thus, its position in the QoS schematic is not fixed.

The establishment of the QoS schematic opens up the possibility of a more analytic investigation of different dimensions contributing to the quality of a service. This is the major issue of the proposed classification. In a first place, the entities allow to analyze which quality dimensions are influenced by the physical transmission characteristics and service parameters. The physical characteristics result in perceptive factors (loudness, articulation, perception of noise or echo, etc.) which have different impact to various quality dimensions, cf. Chapter 3. Then, quality assessment methods are presented which scope at different quality dimensions. It is shown that the differences according to the QoS concept have major consequences for choosing appropriate methodologies in order to obtain valid quality judgments (see Chapter 4). The same holds true for the quality prediction models which are discussed in Chapter 5. New auditory test results (Chapters 6 and 7) underline the necessity for a differentiated view of quality components in order to be able to correctly interpret quality judgments and their estimates.

Chapter 3

FACTORS INFLUENCING THE QUALITY OF SERVICE

There are several possibilities for classifying the factors which influence the quality of a service. As telephone systems are designed to replace a face–to–face conversation, this situation is taken as a reference for the use of a telecommunication service. Factors can be identified which relate to different conversation modalities. The transmission paths which are involved in this type of conversation are illustrated in Figure 3.1. Starting from this schematic, Richards (1973, p. 116) defines factors that cause loss of communication efficiency of the following types:

- Type A: Factors that cause difficulty in listening when the transmission is unidirectional and no assistance is given to the listener by the talker.
- Type B: Factors that cause difficulty in talking.
- Type C: Factors that cause difficulty specifically in conversing, that is, factors associated with the alternation of talking and listening roles of the participants.

This classification, however, is a general one and does not allow an association with physical parts of the transmission system (nevertheless, it gives indications for appropriate assessment methodologies and can easily be associated with quality features of the QoS schematic). A more useful way is to describe factors with regard to their associated perceptive effects. This will be done in Section 1. The factors were chosen in order to describe the perception of signal–related characteristics. In Section 2, these factors are related to basic transmission parameters of the telephone connection. A simulation model is presented which allows the perceptive effects to be investigated in auditory tests, taking the transmission parameters as input values. The development of this model forms a major part of the described work. It is used for all of the tests reported in the following chapters. Finally, transmission parameters are classified according to the QoS schematic given above.

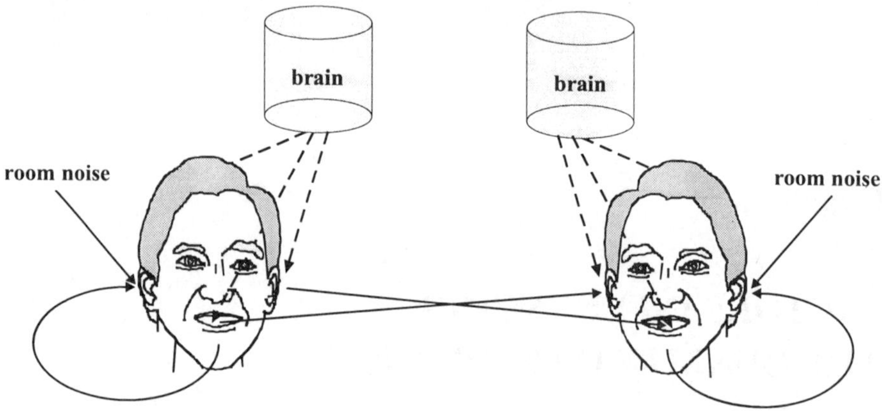

Figure 3.1. Transmission paths involved in a face-to-face conversation.

1. PERCEPTIVE FACTORS

If the face–to–face conversation of Figure 3.1 is to be replaced by a conversation over a communication system, it is of primary importance that the information flow from the mouth of the talker to the ear of the listener is maintained, via an intermediate electrical path. Loudness and articulation are the more important perceptive factors necessary for the information to be transmitted. However, it is not necessary to transmit the speech signal as accurately and completely as possible. A severe limitation which is inherent in telephone transmission is the bandwidth. In narrow–band transmission only the range 300–3400 Hz is generally transmitted, wideband systems use the range 100–7000 Hz. Before the development of medium–to–low–bitrate codecs, the aim was to reproduce the speech signal within the transmitted band as accurately as possible, using waveform codecs. Nowadays, codification algorithms operating at bitrates below 16 kbit/s reproduce a speech signal which is perceptively almost equivalent to the original signal produced by the talker, but requires only a little part of the original information to be transmitted. Apart from these limitations, the quality of the perceived speech signal suffers from noises occurring in transmission as well as from ambient noises. The listener will hear his/her own voice via different sidetone and echo paths. Unwanted echo and overall delay additionally impact the natural conversation flow.

Following these considerations and some empirical evidence given in Section 6 of Chapter 6, the perceptive factors can be categorized as follows:

- loudness (resulting in the concept of loudness rating),
- articulation,

- perception of the effects of bandwidth and linear frequency distortion,
- perception of one's own voice (sidetone),
- perception of echo,
- perception of circuit noise (continuous, impulsive, bursts),
- effects of environmental noise and binaural hearing, and
- effects of delay.

As stated before, the factors were chosen in order to describe the perception of signal–related characteristics of speech. The choice corresponds to the effects resulting from the components of a traditional (analog as well as digital) transmission system. More specific perceptive effects of new (digital) equipment are addressed in Appendix B.

1.1 LOUDNESS

Speech sounds show a continuous, short–time stationary spectrum. The stationary intervals of speech are in the order of approx. 20 ms. However, useful explanations can be drawn from psychoacoustic experiments with stationary sounds. Some basic assumptions are therefore made with reference to stationary, continuous spectrum sounds.

A sound at a single frequency f_k causes a maximum excitation of the nerve fibers on the basilar membrane located at a point of distance l_k from the helicotrema. The excitation due to this single tone can be masked, i.e. the tone can be rendered inaudible, by the simultaneous presentation of a second sound with continuous, constant spectrum, whose frequency spectrum includes f_k. The limit of audibility of the pure tone is reached when the excitation caused by the masking sound, integrated over the relevant length $(\Delta l)_c$ of the basilar membrane, just exceeds the excitation caused by the pure tone at l_k. The critical length $(\Delta l)_c$ is centered at l_k, and is about the same at every location of the basilar membrane. On the other hand, the corresponding critical frequency bands $(\Delta f)_c$ vary with f_k in a non–linear, monotonous way. Results from masking experiments show that sound intensities are integrated within the hearing organ into critical bands with bandwidth $(\Delta f)_c$. Different values for $(\Delta f)_c$ have been determined by Fletcher and by Zwicker (1961). The standardized method for determining loudness levels from instrumentally measurable frequency spectra (ISO R 532 B, 1975; DIN 45631, 1991) is based on $1/3$ octave bands.

The notation of critical bands can be used to determine the loudness level of a continuous spectrum sound, e.g. of speech transmitted over a telephone line. The threshold of audibility of a continuous spectrum sound is, according to

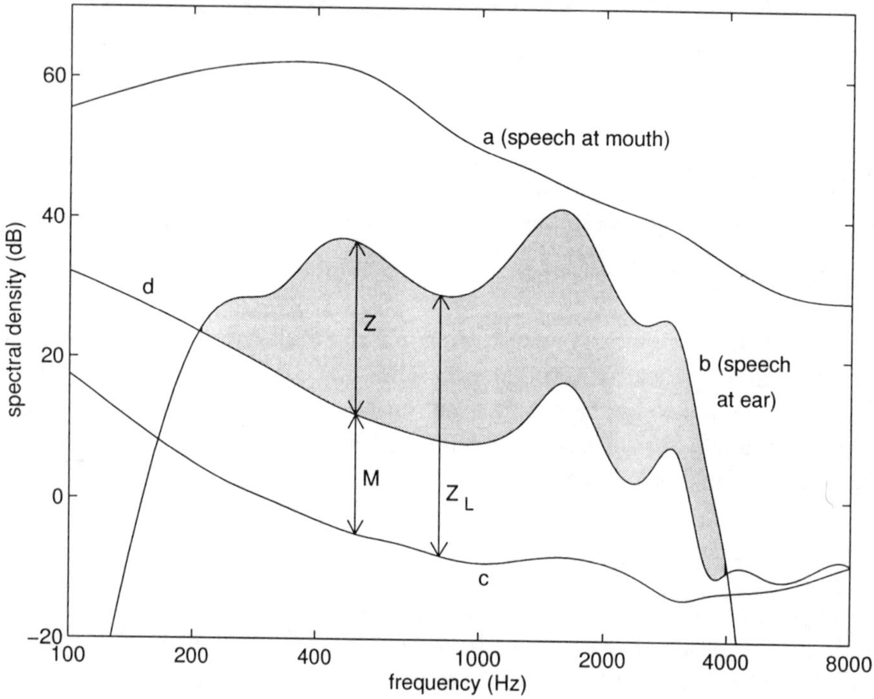

Figure 3.2. Relationship between speech spectrum and threshold of hearing for loudness calculation. a: speech spectrum at the speaker's mouth; b: speech spectrum at the listener's ear; c: hearing threshold for continuous spectrum sounds; d: hearing threshold masked by noise. Numerical values are taken from Richards (1974).

the theory of critical bands initiated by Fletcher, equal to the pure tone threshold β_0 minus an allowance K_C, which is equal to the critical bandwidth expressed in dB, i.e. $K_C = 10 \; log_{10} \; (\Delta f)_c$. In Figure 3.2 the hearing threshold of continuous spectrum sounds (curve c, $\beta_0 - K_C$) is depicted together with the spectrum density of speech at the mouth reference point of the talker, β'_S (curve a). The transmission channel introduces a frequency dependent loss L_{ME} which reduces the spectral density of the emitted sound. The spectral density of the sound reaching the ear of the listener is depicted in curve b of Figure 3.2. In the absence of noise, the loudness of the perceived sound is a function of the interval Z_L between curves b and c.

When noise is present, the threshold of hearing is raised by the (frequency dependent) amount M following the masking effect due to the noise. The parts of the spectrum contributing to the loudness of the perceived sound will then be a function of $Z = Z_L - M$. Z is called the sensation level, and M is a direct function of Z. The loudness of a continuous spectrum sound can now be directly calculated from the contributions Z following a procedure described by Fletcher and Munson (1937). The values can be transformed into quantities $Q(Z)$, which can be seen as the additive entities forming the impression of loudness. Different parts of the frequency range of interest in telephony contribute to differing degrees to the loudness impression, so a weighting factor B' is included in the calculation. B' is derived under the assumption that a given distance interval on the basilar membrane provides a constant weight independent of its location. Therefore,

$$B' = \frac{(\Delta l)_c}{(\Delta f)_c} \cdot \frac{1}{l_{max}} = \frac{0.0156}{10^{K_C/10}} \tag{3.1}$$

The total impression of loudness λ is proportional to the integral over $Q(Z) \cdot B'$, thus

$$\lambda = const \int_0^\infty Q(Z)\, B' df \tag{3.2}$$

The growth function $Q(Z)$ relating Z to loudness is asymptotic to

$$Q(Z) = const \cdot 10^{Z/10} \tag{3.3}$$

for low values of $Z \lesssim 10$ dB, and asymptotic to

$$Q(Z) = const \cdot (10^{Z/10})^{0.3} \tag{3.4}$$

for higher values $Z > 20$ dB.

$Q(Z)$ is depicted in Figure 3.3, which also gives the somewhat differently shaped functions for articulation and the estimation of Y_{LE}. These functions will be discussed later on. The frequency weighting factors B' for loudness are very similar to those derived for articulation or listening–effort opinion Y_{LE}, cf. Figure 3.4. The values presented here are all taken from Richards (1973, 1974). Other sources yield slightly different relations and values. E.g., Zwicker's critical bands (Zwicker, 1961) attribute more relative weight to the lower frequencies than those derived by Fletcher. Zwicker and Fastl (1999) show that Fletcher's underlying assumption of equal power of the pure tone and the masking sound is incorrect. Nevertheless, Richards' values were chosen here because they allow a direct relationship to the principle of 'loudness ratings' used in telephony, and to one of the subjective models presented in Chapter 5.

In telephony, it is normally not necessary to determine absolute loudness levels; it is more common to compare the loudness of one sound to that of another. If absolute loudness levels have to be derived, reference should be made to methods given in DIN 45631 (1991) or ISO R 532 B (1975). The concept

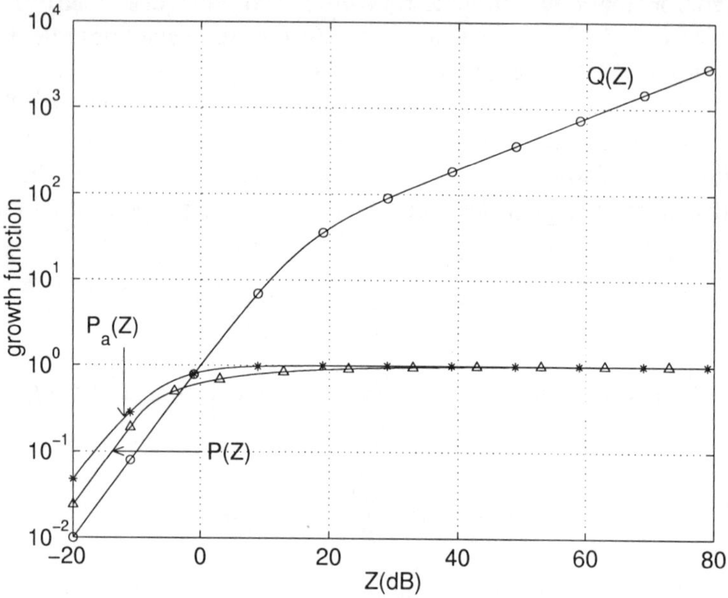

Figure 3.3. Growth functions for loudness, articulation and listening–effort. $P(Z)$: listening–effort (ITU–T Suppl. 3 to P–Series Rec., 1993); $P_a(Z)$: articulation growth function (Collard, 1929); $Q(Z)$: loudness (Richards, 1973).

of loudness ratings is a very fundamental one for the planning and description of analog telephone networks. In principle, a loudness rating, which has the dimension and sign of a loss, is defined by the amount of loss to be inserted into a reference system to secure the equivalence of perceived loudness to the unknown speech path under investigation. Unlike 'reference equivalents' which were used in the past, loudness ratings can be added algebraically for a chain of speech paths, and the reference system used is the so–called 'Intermediate Reference System' (IRS) described in ITU–T Rec. P.48 (1989). They can be derived either by means of auditory experiments using reference systems for comparing perceived loudness (ITU–T Rec. P.78, 1996), or by measuring the sending and receiving sensitivities of the whole transmission path (ITU–T Rec. P.79, 1993). Therefore, it is first necessary to define the acoustic terminals of the transmission. In general, a 'mouth reference point' (MRP) about 25 mm in front of the lips and an 'ear reference point' (ERP) at the entrance of the ear canal are used for this purpose (ITU–T Rec. P.64, 1997, Annex A).

Starting off from Equation 3.2, the loudness of an unknown speech path (λ_U) and a reference speech path (λ_R) can be calculated from their sensation levels Z_U and Z_R. For the loudness growth function, a general form of Equa-

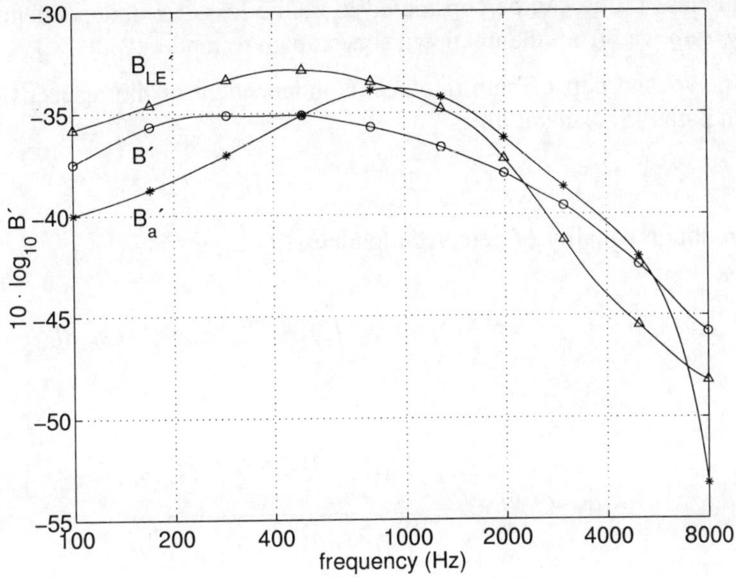

Figure 3.4. Frequency weighting for loudness, articulation and listening–effort. B'_{LE}: listening–effort (Richards, 1974); B'_a: articulation (from Ackroyd, cited in Richards, 1974); B': loudness (Richards, 1973).

tions 3.3 and 3.4 is used:

$$Q(Z) = const \cdot (10^{Z/10})^m \tag{3.5}$$

where a single mean value for m can later be determined by auditory experiments, depending on the range of interest for Z. Let

$$Z_{R0} = \beta'_S - (\beta_0 - K_C) \tag{3.6}$$

so that $Z_U = Z_{R0} - L_{UME}$ and $Z_R = Z_{R0} - L_{RME}$, and use the definition of the sensation level in the absence of noise

$$Z = \beta'_S - L_{ME} - (\beta_0 - K_C) \tag{3.7}$$

Then, by substituting Z in Equation 3.2 one obtains

$$\lambda_U = const \int_0^\infty 10^{(-m/10)\cdot L_{UME}} \cdot 10^{(m/10)\cdot Z_{R0}} \; B' \, df \tag{3.8}$$

$$\lambda_R = const \int_0^\infty 10^{(-m/10)\cdot L_{RME}} \cdot 10^{(m/10)\cdot Z_{R0}} \; B' \, df \tag{3.9}$$

The loudness rating can be considered to be the loss Δx, independent of frequency, removed from the unknown speech path to render $\lambda_U = \lambda_R$.

The second part of both integrals is independent of the respective transmission paths. So, substituting

$$G_L = 10^{(m/10) \cdot Z_{R0}} \; B' \tag{3.10}$$

one can obtain equality of perceived loudness if

$$\int_0^\infty 10^{(-m/10)(L_{UME} - \Delta x)} \; G_L \; df = \int_0^\infty 10^{(-m/10) \cdot L_{RME}} \; G_L \; df \tag{3.11}$$

$$10^{(-m/10) \cdot \Delta x} = \frac{\int_0^\infty 10^{(-m/10) \cdot L_{UME}} \; G_L \; df}{\int_0^\infty 10^{(-m/10) \cdot L_{RME}} \; G_L \; df} \tag{3.12}$$

so

$$\begin{aligned} \Delta x &= \left[-\frac{10}{m} \, log_{10} \int_0^\infty 10^{(-m/10) \cdot L_{UME}} \; G_L \; df \right] \\ &\quad - \left[-\frac{10}{m} \, log_{10} \int_0^\infty 10^{(-m/10) \cdot L_{RME}} \; G_L \; df \right] \end{aligned} \tag{3.13}$$

Without affecting the equality, G_L can be scaled by multiplication with a constant to render $\int G_L df = 1$. G_L can then be regarded as a frequency weighting function. In this way, the loudness rating can be expressed as the difference between the weighted average mouth–to–ear loss of the unknown and the reference speech paths,

$$loudness \; rating = \Delta x = \overline{L_{UME}} - \overline{L_{RME}} \tag{3.14}$$

For computation, the frequency range of interest is divided into a number N of bands. When calculating loudness ratings, normally the 20 ISO–preferred bands are used, spaced at approximately 1/3 octaves from 100 to 8000 Hz. Under this agreement, the weighted average mouth–to–ear loss of the speech path under investigation can be calculated by

$$\overline{L_{UME}} = -\frac{10}{m} \, log_{10} \sum_{i=1}^{N} 10^{(-m/10) \cdot L_{UME,i}} \; G_{L,i} \, \Delta f_i \tag{3.15}$$

Values for $G_{L,i}$ were determined by performing auditory loudness experiments in the CCITT Laboratory. An estimation of m could be calculated from these tests, and the value was $m = 0.175$ for the transmission path. $G_{L,i}$ and $G_{L,i} \cdot \Delta f_i$ can also be tabulated in terms of frequency weights. Substituting

$$W_i = -\frac{10}{m} log_{10} G_{L,i}\, \Delta f_i \tag{3.16}$$

Equation 3.15 can be rewritten as

$$\overline{L_{UME}} = -\frac{10}{m} log_{10} \sum_{i=1}^{N} 10^{(-m/10)(L_{UME,i}+W_i)} \tag{3.17}$$

Tabulated data for W_i can be found in ITU–T Rec. P.79 (1993).

For the definition of loudness ratings, the IRS is taken as the reference system, with an attenuator setting of 0 dB. Having fixed this reference path, L_{RME} becomes constant, i.e. independent of i. Then, Equation 3.14 becomes

$$loudness\ rating = -\frac{10}{m} log_{10} \sum_{i=1}^{N} 10^{(-m/10)(L_{UME,i}-\overline{L_{RME}}+W_i)} \tag{3.18}$$

For a given telephone terminal or transmission path, a value for L_{UME} can be calculated from the measured sensitivities S_{MJ} (mouth–to–junction) and S_{JE} (junction–to–ear), using the measurement set–up described in ITU–T Rec. P.64 (1997). In this measurement, there is a mismatch between the sound pressure measured using an artificial ear and the one existing at a real ear's reference point. Therefore, a correction is recommended to partly compensate for the acoustic leakage. Krebber (1995) developed an artificial ear showing nearly the same leakage characteristics as a real ear, using a simplified pinna model.

Using the sensitivities determined from both the unknown (index U) and the reference (IRS; index R) speech path, definitions of the send (SLR), receive (RLR) and overall (OLR) loudness ratings are obtained:

$$SLR = -\frac{10}{m} log_{10} \sum_{i=1}^{N} 10^{(m/10)(S_{UMJ,i}+S_{RJE,i}+\overline{L_{RME}}-W_i)} \tag{3.19}$$

$$RLR = -\frac{10}{m} log_{10} \sum_{i=1}^{N} 10^{(m/10)(S_{RMJ,i}+S_{UJE,i}+\overline{L_{RME}}-W_i)} \tag{3.20}$$

$$OLR = -\frac{10}{m} log_{10} \sum_{i=1}^{N} 10^{(m/10)(S_{UMJ,i}+S_{UJE,i}+\overline{L_{RME}}-W_i)} \tag{3.21}$$

with the factor $m = 0.175$. In loudness rating calculations for network planning purposes, only the band 200–4000 Hz is generally taken into account. To make allowances for the differences to wideband measurements, the according weighting factors are then decreased by 0.3 dB. Loudness ratings determined in this way show a relatively good additivity property, i.e. the loudness rating of a transmission path can be determined by simply adding together the loudness ratings of the constituting parts (in dB). The description of the speech transmission path in terms of loudness ratings will be discussed further in Section 2.

1.2 ARTICULATION

Articulation and intelligibility measurements are sometimes used as a measure for the information capacity of a speech transmission path. Both give the percentage of units of speech material presented in a listening–only experiment which is correctly identified. In general, the term 'intelligibility' is used for complete and meaningful words, phrases or sentences as speech units, and the term 'articulation' for word fragments, syllables, phonemes or meaningless words. Thus, 'articulation' refers to the correct identification of the form only and 'intelligibility' to the correct identification of the contents by means of the form. However, both investigate signal–related characteristics. From the user's point of view, articulation is a prerequisite for 'comprehensibility' and intelligibility for 'comprehension'. Comprehension involves not only the phonetic–linguistic aspects but also semantic and pragmatic knowledge. It presupposes comprehensibility, i.e. the possibility of understanding, and the ability and the willingness of the listener to do so (Jekosch, 1998, p. 116). To be coherent with the other perceptive factors only articulation will be discussed here.

Though the effectiveness of the speech path cannot be completely described in terms of articulation scores, the information capacity provides useful pieces of information on the performance of the whole speech path. The efficiency of the communication will depend on both the characteristics of the speech material and the relevant characteristics of the talker and listener. Measurements on articulation will therefore be influenced by variations between talkers and between listeners, and very strongly by the choice of the assessment method and the speech material used. Jekosch (1998) shows how, apart from talker and listener variability, the test vocabulary, stimulus presentation and scaling method directly take influence on the identification scores. A large number of different test methods have been extensively studied in the past, and a choice of such methods will be discussed in Section 2 of Chapter 4. In order to provide a defined equipment set–up, a reference system called SRAEN used to be recommended.

Similar to loudness, articulation has been the object of modeling efforts since the assessment of telephone connections began. A first theoretical model was developed by Collard (1929), which is very similar to the loudness model presented in Section 1.1. It has been further elaborated by others, but the main features remained unchanged. The aim is to reproduce the results of articulation tests using psychoacoustic features and physical parameters of the transmission path. The psychoacoustic relations have already been described above. Similar to loudness, a quantity called "band articulation" is calculated from the sensation level:

$$band\ articulation = \int_0^{\infty} P_a(Z)\ B'_a df \qquad (3.22)$$

$P_a(Z)$ is an articulation growth function analogous to $Q(Z)$ used in loudness, and B'_a is a frequency weighting function. $P_a(Z)$ is defined by

$$P_a(Z) = \frac{10^{Z/10}}{10^{Z/10} + 0.2} \qquad (3.23)$$

The frequency weighting function B'_a differs slightly from the one used for calculating loudness in that it puts more emphasis on the center frequencies and less on the upper and lower bands. Both $P_a(Z)$ and B'_a are depicted in Figures 3.3 and 3.4, together with the according functions for loudness. Band articulation is not the same as an articulation score which could be measured in one specific test. However, for one specific test, a unique function can be derived which establishes this relation:

$$articulation\ score = \Phi(band\ articulation). \qquad (3.24)$$

An example of a function Φ relating band articulation to scores from English logatom tests using a well–trained crew and a standardized reference system is given by Richards (1973, p. 172).

1.3 EFFECTS OF BANDWIDTH RESTRICTION AND FREQUENCY DISTORTION

Linear distortions due to the frequency characteristics of the transmission channel may have a strong influence on the overall perceived quality of a telephone connection. Up to now, nearly all telephone connections have been limited to the narrow-band case, i.e. 300–3400 Hz. This frequency band is imposed by the channel filter defined in ITU–T Rec. G.712 (1992). The attenuation/frequency distortion as well as group delay slightly depend on the type of channel and port (analog/digital, 2–wire/4–wire). Greater differences result from the terminal equipment (handset shape and electrical characteristics, earcap leakage) and from low–bitrate codecs. Unfortunately, the effect of linear distortion has only rarely been the subject of perceptive experiments.

Gleiss (1970) reports a couple of paired comparison experiments with speech samples differing in the upper and lower limit of the transmitted frequency band. The entity to be assessed was naturalness. It was found that naturalness is much more affected by a shift in the lower band limit than in the upper band limit. E.g., a 100 Hz band at the lower limit is approximately equivalent to a 5–10 times larger band at the upper end. This relationship is approximately the same as the one between the corresponding critical bandwidths in hearing. Furthermore, the preference judgment for naturalness is essentially invariant to noise and speech levels, but variable with the talker characteristics. Another factor affecting the naturalness judgment is the listener's familiarity with voices. On the other hand, the importance of bandwidth on naturalness is not identical to the corresponding effect on intelligibility. The investigations cited in Section 6 of Chapter 6 underline that these dimensions are mainly uncorrelated. Work performed by Thomas (1968) suggests that naturalness may depend more on the first formant and intelligibility more on the second formant. In any case, it is justified to rate these factors independently, or to use appropriate multidimensional methods. Further experimental data on different frequency bandwidths will be given in Section 4 of Chapter 7.

Within the transmission band, the preference for a flat or sloped frequency response depends on the upper cut–off frequency (Gleiss, 1970). For a more narrow band, a positive slope is preferred. In general, a recommended range of response curves would cover slopes between 0 and +3 dB/octave (Gleiss, 1989). Experiments reported by Bücklein (1962) on wideband speech and music signals show that peaks in the frequency response are much more annoying than corresponding holes. This observation refers to both detectability and intelligibility. Experiments on wideband speech and music by Gabrielsson et al. (1985) allow single perceptive attributes to be related to characteristics in the frequency response of a sound reproducing system. "Fullness" ("full" in opposition to "thin") and "softness" ("soft and gentle" in opposition to "sharp, hard, keen and shrill") seem to depend very much on low (< 1 kHz) frequency contributions, whereas "brightness" ("bright" in opposition to "dull and dark") is more dependent on high frequency bands. A steeply rising frequency response also contributes to a perception of "sharpness" (opposite to "softness").

1.4 PERCEPTION OF SIDETONE

A telephone user who is engaged in a conversation using a handset will hear his/her own voice via several mechanical, acoustic and electrical paths. First, the free ear, not shielded by the handset, will pick up a certain amount of speech as well as ambient noise. In telephony, it is generally assumed that this path is more or less "switched off" by central masking of the human brain. Nevertheless, this assumption is doubtful at least for higher levels of ambient room noise. A second and more important sidetone path is the acoustic one to the engaged ear, i.e. where the handset is held. It is mainly due to bone conduction in the user's head, and to the leakage between the handset and the human ear. Also a mechanical path via the handset exists. A third sidetone path is an electrical one through the handset itself. This path used to be caused by an impedance mismatch between the subscriber line impedance and the so–called sidetone balance impedance of the telephone set. Nowadays it is introduced by different electric circuits in a defined way.

In telephony classification of sidetone is made in terms of 'talker sidetone' and 'listener sidetone'. Talker sidetone includes all the paths contributing to one's own voice transmission. All the paths contributing to the ambient noise transmission to the engaged ear are classified as listener sidetone. There are several good but also several inconvenient points of having sidetone. In the ideal case, the talker sidetone path would be the one which provides an acoustic signal identical to the one perceived when speaking without any handset, in a free conversation (Williams and Barnes, 1974). However, this would require a capsule with a low acoustic impedance, and a large acoustic leakage which is impractical in normal telephone situations. The talker sidetone path is also important because it gives a feedback to the talker informing him/her that the apparatus is working. This is regarded as the most important feature of sidetone. On the other hand, a loud electrical sidetone will cause major problems when room noise is present. It will further encourage the talker to speak more quietly, which is a problem in high loss connections. Therefore, a relatively high electrical loss is required for the sidetone path.

The sensitivity of the sidetone path as a function of the frequency can be measured, as described in ITU–T Rec. P.64 (1997). This sensitivity is different for the directed sound from the (artificial) mouth (S_{meST}) and for the diffuse sound resulting from ambient room noise (S_{RNST}). The (frequency dependent) difference between these sensitivities is termed

$$DELSM = S_{RNST} - S_{meST} \tag{3.25}$$

From these sensitivities, loudness ratings can be calculated, similarly to those described in Section 1.1: the sidetone masking rating ($STMR$) for talker sidetone and the listener sidetone rating ($LSTR$). If one compares the frequency responses between electrical and acoustic sidetone paths, it can be seen that the attenuation of the acoustic paths (due to acoustic leakage and bone conduction) is relatively small for low frequencies, but is considerable for high frequencies (ETSI Technical Report ETR 250, 1996, p. 46). In other words, the acoustic sidetone path dominates over the electrical one for frequencies below approx. 600–800 Hz. In the higher frequency region, the electrical sidetone path is important for preserving good conversational conditions in high ambient noise conditions (ITU–T Rec. P.76, 1989). Therefore, more emphasis is put on the high frequency areas of the electrical sidetone path than this is done for SLR, RLR and OLR. In Table 3.1 a comparison is made between the weighting factors for OLR, for $STMR$ and for the echo loss, Le. Weights are expressed in terms of K_i, when Equation 3.18 is expressed by

$$loudness\ rating = L_0 - \frac{10}{m} log_{10} \left(\sum_{i=1}^{N} K_i \cdot 10^{(-m/10) \cdot L_i} \right) \tag{3.26}$$

where L_i is the loss of the path under consideration, L_0 is a constant which is dependent on what loudness rating is to be computed, and

$$K_i = 10^{(-m/10)(W_i - L_0)} \tag{3.27}$$

It can be seen that for both sidetone and echo paths lower frequency components are neglected. For the sidetone path, an emphasis is put on the bands above 1 kHz, while the weighting of the echo path is nearly constant in the transmitted band. Thus, for talker echo, contributions from all the frequency bands are relevant, while for sidetone, only the upper frequency region is of importance.

The loudness growth factor m for sidetone loudness calculations is slightly higher ($m = 0.225$) than for OLR, SLR and RLR ($m = 0.175$). It generally depends on the sound level at which the loudness comparison is assumed to be made. For even lower levels, $m = 0.5$ corresponding to voltage addition, or $m = 1.0$ for very low sounds at the threshold of hearing, e.g. for echo loudness ratings (corresponding to power addition). For small values of m and for a moderate spread in the L_i values, Equation 3.26 can be simplified to

$$loudness\ rating = L_0 + \sum_{i=1}^{N} K_i \cdot L_i \tag{3.28}$$

Under the condition of validity of Equation 3.28, the total loudness rating of a connection can be computed by simply adding together the loudness ratings of its separate parts. This is the underlying assumption for many of the transmission rating models discussed in Chapter 5.

1.5 PERCEPTION OF ECHO

Echo occurs in all modern telephone networks where a considerable transmission delay is introduced by the equipment. It mainly results from a coupling between the go and return path, used for converting between 2–wire and 4–wire

Table 3.1. Comparison of frequency weighting functions for calculating loudness ratings. Values for OLR calculated from Tab. A.2 of ITU–T Rec. P.79 (1993) using Equation 3.27 with $L_0 = 0$, values for $STMR$ and Le taken from Tab. A.1 of ETSI Technical Report ETR 250 (1996).

mid frequency (Hz)	$K_i(OLR)$	$K_i(STMR)$	$K_i(Le)$
100	0.013	0	0
125	0.040	0	0
160	0.071	0	0
200	0.070	0	0
250	0.086	0.01	0
315	0.063	0.02	0.05
400	0.106	0.03	0.10
500	0.067	0.04	0.10
630	0.078	0.05	0.10
800	0.078	0.08	0.10
1000	0.072	0.12	0.10
1250	0.052	0.12	0.10
1600	0.059	0.12	0.10
2000	0.037	0.12	0.10
2500	0.042	0.12	0.10
3150	0.021	0.12	0.05
4000	0.045	0.05	0
5000	0.024	0	0
6300	0.024	0	0
8000	0.045	0	0

parts (the so–called hybrids). The degree of mismatch between the impedances determines the amount of the signal reflected, and thus the loudness of the echo. Echo can be categorized into talker echo (if the talker is affected by his own signal) and listener echo (resulting from multiple reflections). Listener echo results in a more 'hollow' sound when the round trip delay is low. Both the talker and the listener echo can be described in terms of the echo path delay and the loudness of the echo. The latter is modeled using a loudness rating, called talker echo loudness rating ($TELR$) for the talker echo, and the weighted echo path loss ($WEPL$) for the listener echo. $TELR$ is equivalent to the sum of the sending loudness rating (SLR), receiving loudness rating (RLR) and the weighted loss of the echo path, Le. The frequency weighting for $TELR$ has already been discussed above. For the listener echo, in general an addition on voltage basis is done ($m = 0.5$) on a linear frequency scale. However, calculation on a logarithmic scale yields very similar results. It should be noted that listener echo is not very important in properly designed networks.

The disturbance due to talker echo will be discussed further in Section 1 of Chapter 7. It should be noted that for very low echo path delay times (e.g. $T < 10$ ms), the echo will be interpreted as talker sidetone. In such cases, a masking will occur between the signal resulting from the echo path and the sidetone signal. ETSI Technical Report ETR 250 (1996) gives some examples of the impact of low–delay echoes, sometimes called 'delayed sidetone'.

1.6 PERCEPTION OF CIRCUIT NOISE

Various sources of noise occur in telephone networks. In analog frequency division multiplex networks, noise was one of the limiting factors of long–distance connections. With the introduction of digital circuits, the situation has changed. Circuit noise is much less of a problem. However, mobile communication enforces much higher levels of ambient noise, a degradation which is amplified by the worse acoustic situation of several modern, straight handsets. Special devices such as low–bitrate codecs, voice activity detectors, comfort noise injection, etc. introduce especially non–stationary impairments which can, in the first instance, be subsumed under the term "noise". The effects of an erroneous transmission, resulting in random and burst errors, are often modeled as "noise" as well. Many of these impairments have not yet been subject to a more detailed experimental investigation. Different from additive noise is the so–called signal correlated noise, mainly caused by quantizing distortion. It has been extensively studied, and a laboratory system which introduces such distortions (the 'modulated noise reference unit', MNRU) is currently being recommended as a reference in auditory tests on speech codecs (ITU–T Rec. P.810, 1996). The MNRU and its related effects will be discussed later in Section 2 of Chapter 6.

As a general rule, the perceptive effect of noise depends on the power, on the frequency distribution and on the amplitude distribution of the noise. The loudness of wideband circuit noise is generally characterized by the level measured using a so–called psophometer as recommended by the ITU (ITU–T Rec. O.41, 1994). The frequency weighting introduced by the psophometer is similar to the C–message weighting used in the US. ITU–T Rec. P.11 (1993) states that psophometrically measured levels also satisfactorily characterize the effect of induced power hum, as experimental data show. On the other hand, impulsive noise and especially long–term time–variant noise (caused by noise bursts) cannot be sufficiently characterized by their psophometrically measured levels. For impulsive noise, where the excursion of the noise waveform on a channel exceeds a specified level threshold above the rms value of the noise, the occurrence and amplitude distribution of the individual impulses are of importance. Automatic classifiers have been proposed which can discriminate between stationary and impulsive noise using signal processing techniques (ITU–T Delayed Contribution D.014, 1997). It still remains unclear to what extent the subjective effect of impulsive noise can be correlated with the occurrence and amplitude of arriving pulses.

1.7 PERCEPTION OF AMBIENT NOISE

Ambient room noise affects the efficiency of the communication in several ways. On the listener side, the room noise reaches the engaged ear through the leakage between the handset and the ear, and via the electric sidetone path. The free ear is far more affected, but it is assumed that central masking processes and the attention of the listener towards the engaged ear will lead to this path being nearly "switched off". This assumption may not be satisfied if high levels of room noise are present, or if the room noise has a certain significance to the listener and therefore carries information. Bodden and Jekosch (1996) investigated the impact of several perceptively different types of room noise on the overall quality in a listening–only situation. The noise types differed in their spectral and temporal composition, in their information content, as well as in their emotional effect on the listener, e.g. white noise, noise from a railway station, a weeping baby, a pneumatic hammer, music, etc. They found that, though the difference in integral quality judgment is mainly statistically not significant, major differences exist between the auditory judgments (up to one category on a 5+2–point scale as in Figure 4.10) for high levels of room noise . Music was rated best, whereas the railway station type noise had the strongest impact on the judged overall quality. This correlates with findings in the investigation of product sounds, where the informative and emotional contents of sounds turned out to be of major importance to judgments on levels of annoyance (Bednarzyk, 1999).

On the send side, the room noise causes an impact on the transmitted speech signal. It is superposed on it with the transfer function of the diffuse sending characteristic of the handset. Some of this noise is perceptively compensated for by the so–called Lombard effect. This term describes the psychoacoustic effect created when the talker raises his/her voice in a noisy environment, and speaks louder and more clearly. The effect was first described by Lombard in 1910, and has been the subject of quantitative investigations by several authors (Lane et al., 1961, 1970). A common result of these studies is that the speaker's effort to compensate for the reduction in signal–to–noise ratio typically results in only a halfway compensation, when measured in decibel units. The common explanation for this is that the exponent of the so–called autophonic loudness scale (the speaker's estimation of his/her own vocal level) is around twice the corresponding exponent of the sone loudness scale (the normal loudness impression for ambient noise). In this way the speaker matches the voice in order to keep the signal–to–noise ratio approximately constant, but he/she is mislead by the perceptive loudness characteristics, and compensates only half the amount. The Lombard effect is not taken into account in some of the models for predicting perceived overall quality. It can only be investigated accurately in natural conversation situations, because both the talker and listener will adapt their speaking style to the room noise situation in their own as well as in the other person's room. Further experimental results on room noise will be discussed in Section 1 of Chapter 7.

1.8 EFFECTS OF TRANSMISSION DELAY

Besides causing echo when reflections occur, absolute delay, i.e. the overall delay between the sound emitted by the talker and the one reaching the listener's ear, can cause difficulties in conversation. Within a dialog, one conversation partner expects a response from the other within a certain time window. The width of the window depends on talkspurt (the length of the previous utterance), and utterance speed (Kitawaki and Itoh, 1990, 1991). The latter varies according to the individual and to the topic of conversation. When the response of the dialog partner does not correspond to the expected window due to delay, this is normally not directly attributed to the connection, but in the first instance to the conversation partner. Thus, even major degradation due to waiting and double–talk cause a decrease in task performance and in the so-called "conversation efficiency", but they are not necessarily rated as a quality degradation of the system or service. This observation described by Karis (1991) has been verified in experiments carried out to verify the E–model, cf. Section 1 of Chapter 7.

The impact of absolute delay is largely dependent on the conversation task. Kitawaki and Itoh (1990) showed that the detectability threshold varied between 100 and 700 ms for trained test crews, and between 350 and 1100 ms for inexperienced subjects, depending on the conversation scenario. The tasks were e.g. to read aloud random numbers as quickly as possible, but also to converse freely. These large differences make it clear that for assessing quality in 'normal' situations, the results of experiments using special tasks or highly trained subjects (Kitawaki and Itoh, 1991) are doubtful. They can help to clarify how delay affects the run of a conversation, but not how these effects lead to quality judgments. The test results from Kitawaki and Itoh (1991) also show that, even if the delay is not detected by the subjects, it can indirectly have an influence on the quality ratings (comparison of Fig. 7 and 8 in Kitawaki and Itoh, 1991).

One interesting effect was observed in a Bellcore experiment (ITU–T Contribution COM 12–62, 1990). The overall quality judgment rose although delay times increased in the range $Ta > 600$ ms. One explanation for this may be that the user's behavior changes in response to the long (and detectable) delay. It can also be observed when talking to someone who is expected to understand and follow a dialog more slowly, e.g. to a foreigner.

2. CONFIGURATION OF A TELEPHONE CONNECTION

Besides the perceptive factors, it is useful to have a look at the network components which form a complete telephone connection, because these quality elements are in the hands of the planner in order to set up high–quality networks. Generally, they can be divided into three main groups: terminal elements, connection elements, and transmission elements. The terminal elements most commonly used are all types of analog or digital telephones, either wirebound or mobile. All types of switching equipment (analog or digital) are subsumed under 'connection elements'. The transmission elements are the media between the connection elements which finally transmit the information.

There are many different potential network configurations with which the network components can be combined. For modern networks it is often impossible to define a configuration in advance which will be encountered for a specific connection, due to interconnection between different network operators. However, it is possible to describe the equipment in use by defining the associated effects on the transmission quality. The effects are related to one or several pieces of equipment and can be categorized as shown in Figure 3.5. This categorization is more convenient than the one cited above and will be used as a further basis of discussion. It also corresponds to the input parameters of some network planning models, especially to the ones of the E–model (cf. Chapter 5).

Most telephone connections are nowadays terminated by a wirebound or mobile handset. The handset can be characterized by the sending and receiving sensitivity, the sidetone characteristic, and its sensitivity to diffuse sound (all as a function of frequency). Electro–acoustic sensitivities are measured between the mouth reference point (MRP) or the ear reference point (ERP) and the line, which should be terminated by a 600 Ω resistance. The measurement principle as well as the location of the reference points are described in ITU–T Rec. P.64 (1997). For network planning purposes, the measured frequency characteristics are usually combined forming a weighted scalar value, namely the loudness rating, as defined in Section 1.1. In terms of loudness ratings, the telephone set can be described by the send loudness rating SLR, the receive loudness rating RLR, and the sidetone masking rating $STMR$. $STMR$ is

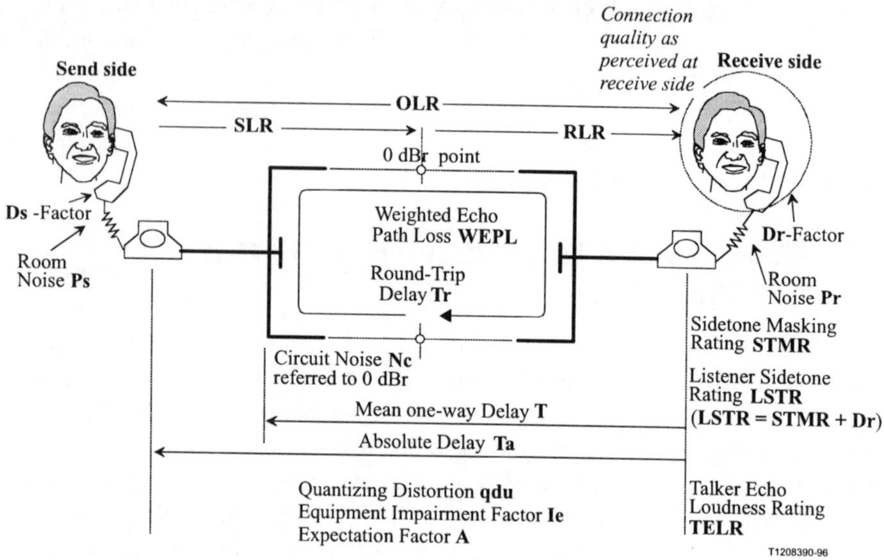

Figure 3.5. Basic configuration of a telephone connection (ITU–T Rec. G.107, 1998, Figure 1). The parameters are discussed in the text of this section and of Chapters 6 and 7.

a function of the terminal related parts of SLR and RLR, as well as of the sidetone path loss Lst. The sensitivity for listener sidetone, i.e. to the room noise, is expressed by the listener sidetone rating, $LSTR$. For talker sidetone, the transmitting sensitivity relates to the direct sound (emitted from an artificial mouth), whereas the according sensitivity for the listener sidetone relates to the diffuse sound. The difference between these sensitivities is termed $DELSM$, and a frequency weighted version of $DELSM$ is the so–called D–factor. The D–factor can also be expressed by

$$D = LSTR - STMR \tag{3.29}$$

D depends only on the shape of the handset and the microphone circuitry, and may (for non–linear microphones) vary with the speech level.

Apart from the terminal equipment, circuit elements contribute to the overall transmission characteristic between MRP and ERP. However, frequently in network planning a virtual reference point in the middle of the connection is defined, the so–called 0 dBr point (because line levels refer to this point). Loudness ratings are then defined between MRP or ERP and the 0 dBr point, thus

$$OLR = SLR + RLR \tag{3.30}$$

The OLR frequency characteristic is mainly a BP filter response of the transmitted band (300–3400 Hz), but shaped with the SLR and RLR characteristics.

Signal reflections occur at the coupling between the go and return path in the conversion between 2–wire and 4–wire circuits, the hybrids. In association with considerable delays (which are increasingly likely in modern networks) they can result in noticeable echo. The echo in the direction of the talker can be characterized by its loudness and by the mean one–way delay time, T. The loudness is normally characterized by the associated loudness rating, the talker echo loudness rating $TELR$. If multiple reflections occur, the listener echo might also be annoying. Listener echo is described by the so–called weighted echo path loss ($WEPL$) and the delay Tr. In addition to annoyance, low values for $WEPL$ increase the risk of instability ('singing'). Apart from echo, a long one–way transmission time can disturb the communication *per se*, thus causing difficulties in the normal flow of turns. The influencing parameter is the absolute delay, Ta.

Various noises occur in telephone networks. Electric circuit noise, which was a limiting factor in traditional analog networks, has become less important since the introduction of digital circuits. All the noises introduced by the various circuits are power summed and described by a unique parameter Nc, which refers in level to the virtual 0 dBr point. Induced noise on subscriber lines, e.g. hum from power and fraction lines, is another source of noise. This is described by the so–called noise floor parameter, $Nfor$. A more limiting parameter, especially in mobile communication, is the ambient room noise. Room noise reaches the listener directly via the free ear, and the engaged ear via the leakage between the handset and the listener's ear. The room noise is also transmitted via the sidetone path, and it is transmitted from the talker to the listener. This latter effect of room noise can be treated as an additional circuit noise, and this is done in some of the models for quality prediction discussed in Chapter 5. Room noise is mostly characterized by its A–weighted equivalent continuous sound pressure level, either in the talker (Ps) or in the listener (Pr) environment. Quantizing distortion from logarithmic PCM systems results in a signal–correlated noise. For PCM systems used in telephony, it is described either by the signal–to–quantizing–noise ratio Q, or by the so–called quantizing distortion unit, qdu. One qdu is defined as the degradation introduced by one logarithmic PCM coder–decoder pair. A reference unit producing signal–correlated noise in an adjustable way is the modulated noise reference unit, MNRU, as described in ITU–T Rec. P.810 (1996). It is used as a reference degradation for auditory tests, cf. Section 2 of Chapter 6.

Signal compression is necessary for reducing bandwidth in modern digital networks. The codecs which are in use in telephony operate at a medium–to–low bitrate between 32 kbit/s for ADPCM coding to approximately 5.6 kbit/s for the GSM-HR codec. Especially codecs operating at lower bitrate introduce impairments which are not comparable to quantizing distortion. The speech sounds more artificial, and the naturalness and speaker recognizability might be affected. Due to its special characteristic the degradation resulting from codecs cannot be characterized by a single parameter. Nevertheless, an attempt was made to perform this using the E–model, including a new 'equipment impairment factor'. Under what circumstances the results obtained in this way are valid will be discussed in further detail in Section 3 of Chapter 7.

3. SIMULATION OF A TELEPHONE CONNECTION FOR QUALITY ASSESSMENT PURPOSES

In the preceding sections, perceptively relevant factors which have an influence on speech communication quality as well as the related network components have been presented. In order to assess the perceived quality an actual user experiences in conversing over a connection, it is important to have a tool at hand which implements all the relevant transmission paths with the according physical characteristics. The assessment results can then be used to make predictions concerning expected quality, and to investigate quality prediction models. Such a tool should be highly flexibile, so that it can be used for different tasks and various network configurations. A simulation tool seems to be an appropriate solution. Consequently, a telephone line simulation has been designed and implemented. It is used in the following for the development of assessment methods (Chapter 4) and for all auditory experiments described in Chapters 6 and 7.

The following requirements should be met by the simulation:

- It should implement a realistic reproduction of the telephone connection, including all the relevant transmission paths (even sidetone and echo). However, making use of real life equipment, which would offer far less flexibility in manipulating the relevant parameters, is not desirable and would be expensive.

- All the parameters of the transmission (cf. Figure 3.5) should be adjustable within reasonable limits. The manipulation should be fast and easy to use, in order to simulate connections with different transmission characteristics within one auditory test session.

- In order to be able to carry out conversation experiments (see Section 3 of Chapter 4), the transmission should work in real time and should respect the delays encountered in networks.

- The system should, if necessary, be flexible enough to include future impairments or equipment with special transmission characteristics as easily as possible.

- The simulation should work in listening–only and conversation scenarios, using the same transmission characteristics.

- It is important that the terminal which is provided for the assessing person matches the one which is encountered in a real application. I.e., for the test subject as well as for a potential conversation partner normal standard handsets or, if desired, headsets or hands–free terminals should be provided. It should be possible to connect different terminal elements to the simulation.

It is obvious that not all the requirements can be given the same priority. For the purpose of this work, it seems appropriate to concentrate first on traditional impairments which are especially encountered in wirebound (analog or digital) networks. In order to be able to investigate the quality predictions made by the models discussed in Chapter 5, at least all the input parameters to such models have to be adjustable in a defined way.

The simulation working in real time can be used to directly compare different assessment methodologies, e.g. conversation and listening–only tests. Especially in test set–ups enabling a normal conversation to take place, several characteristics are handled in the same way as in a normal telephone conversation, namely

- the variation in speech level,

- the variability in speaking style,

- the variation in the transfer characteristic between mouth and microphone or between microphone and ear,

- the pick–up of ambient room noise by the microphone and by the human listener, and

- the effects of binaural hearing.

Following the requirements set out above, the easiest, cheapest and most elegant way was to set up a software and hardware simulation of the whole transmission to characterize a general telephone connection. The system choosen was a signal processing hardware which can be wired and programmed via software. Software manipulation makes the system more flexible, while the hardware allows the simulation in real or close–to–real time. The structure of the system is depicted in Figure 3.6. The triangles represent filters or programmable attenuators, the rectangles delay lines (for T, Ta and Tr), external codecs and the channel BP filter. It can be seen that all the necessary speech paths (transmission, sidetone, talker and listener echo) are modeled. A difference to real–life networks is the simulation of the talker echo path: whereas the talker

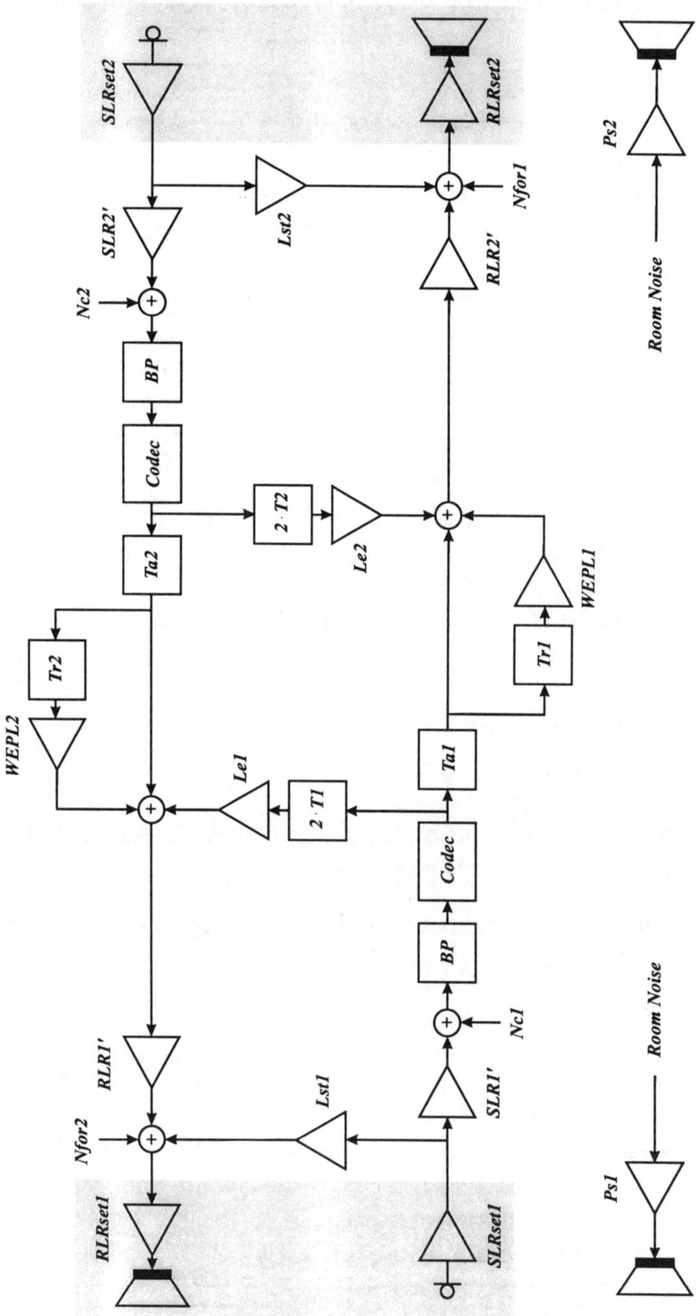

Figure 3.6. Configuration of the telephone line simulator used in the experiments.

echo passes through two coding–decoding processes (because reflections generally occur at the far end hybrid), the simulation only takes into account one coding–decoding pair. This allowance was made for stability reasons, due to the feedback loop which otherwise would have been formed by the two echo paths. In this way, it becomes possible to adjust T and Ta independently, if only one parameter is investigated. The simulation works in full–duplex mode, as all transmission paths are modeled in both directions. All traditional transmission impairments are simulated in such a way that makes use of the description of such impairments for network planning purposes. The transmission characteristics of the relevant paths are described in terms of loudness ratings, while the filter (depicted by the triangle) can be used to produce different frequency characteristics and attenuations for the path under consideration. The noise sources (Nc and $Nfor$) produce white noise, which is psophometrically weighted to calculate the noise level in [dBm0p].

The implementation of codecs poses a problem for general–purpose hardware. Therefore, a commercially available specialized hardware is used, which implements several of the low–bitrate codecs of interest in telephony. The simulation is done by a direct coding–decoding process. Thus, it is not possible to manipulate the encoded data stream, e.g. to include bit errors. Asynchronous tandeming of several codecs is realized on a chain of these DSP cards. A variety of ITU–T codecs can be implemented (some of them at different bitrates), as well as the North American cellular standard and codecs operating at very low bitrates (LPC–10 and MELP). The hardware also allows a modulated noise reference unit (MNRU), used as a reference degradation in some of the tests, to be simulated. This implementation satisfies the description in ITU–T Rec. P.810 (1996) when used in listening–only scenarios. In conversation tests the imposed signal–to–noise ratio is fixed, so that the injected noise level varies in relation to the speech level.

The simulation model is connected to two conventional handsets with good acoustic characteristics, which have been in use in Germany for years ('Type 7' handsets). Both handsets are equipped with piezoelectric microphone capsules and loudspeakers, which are normally used in this type of handset. Microphone and loudspeaker signals are directly connected to the simulator, so that the sidetone path can be controlled. For experiments with mobile telephones, a mobile handset (type 'Sinus 43') was fit with an electret microphone (type 'Sennheiser KE 4') and a typical loudspeaker. In this case as well, microphone and loudspeaker signals are directly coupled to the simulator, using a thin cable connection which avoids degradations inherent in wireless transmission. The usage of a wired handset has to be taken into account for the assessment results. All handsets were acoustically measured (ITU–T Rec. P.64, 1997) regarding their sensitivities in the send and receive direction, and regarding the $DELSM$ parameter. From these measures send and receive loudness ratings and the *D*–

factor were determined. The filter characteristics at the input and output port of the simulator were chosen in order to have for the whole cascade $SLR{*}SLRset$ and $RLR{*}RLRset$ a characteristic corresponding to an intermediate reference system, as described in ITU–T Rec. P.48 (1989). However, the simulator also allows the transfer function to be modified in order to match standards for 'generally encountered characteristics' or for any other desired frequency shape (a variety of frequency characteristics measured on 25 handsets approved for use in Sweden can be found e.g. in the ETSI Technical Report ETR 250 (1996, pp. 38–39).

It is important to note that conventional handsets were used for all the experiments, and not headphones. Dimolitsas and Ravishankar (1995) observed that subject ratings in single–stimulus opinion judgments differ significantly for handset and headphone presentation. The MOS values (mean opinion score, cf. Chapter 3.1) are consistently lower for headphone presentation. The ordinal rank of the codecs under test, on the other hand, was highly correlated for both presentation conditions. This can be seen as an indication that the presentation mode can play a major role, especially when 'absolute' evaluation results are required.

In order to avoid excessive clipping it is important to ensure that the average speech level on the line corresponds approximately to the level encountered in real–life networks. Information in ETSI Technical Report ETR 250 (1996, p. 67) states that on average this level is approximately $-11 - SLR$ [dB] at the 0 dBr point. The assumption was made that in digital networks the overload point of the system corresponds to 6 dBr, resulting in a reasonable average active speech level of approx. 25 dB below the clipping point. However, as the speech level varies during a conversation and between test subjects, no fixed level can be guaranteed and only on average will the level requirements be met.

The simulation model described above was designed and implemented by the author as the principal tool for carrying out auditory tests. Apart from the transmission simulation, a complete test environment was set up in order to to facilitate planning, administration and analysis of the experiments. For the conversational and listening–only experiments two adjacent office rooms were available. These rooms were treated in order to satisfy the background noise conditions given by NC25 (Beranek, 1971, p. 564-566), corresponding to a noise floor of below 35 dB(A). Reverberation time was between 0.37 and 0.50 s in the frequency range of interest. The wall between the rooms was reinforced resulting in a weighted sound reduction index of $Rw' = 45$ dB. These rooms fulfilled all the requirements for test cabinets given in ITU–T Rec. P.800 (1996). Monitoring equipment was provided for the communication between the test cabinets and the experimenter's control room where the simulator was located (intercommunication system). Tape recordings (DAT) were made of at least the two input (microphone) signals, sometimes (for control reasons) also of the two output (loudspeaker) signals. Signaling equipment was simplified omitting the dialing procedure, in order to minimize the effort. It is expected that the lack of naturalness in the call set–up will not have a special impact on the quality opinion of the transmission. However, for assessing acceptability, more effort is required to generate a more realistic and natural situation.

4. CLASSIFICATION OF TRANSMISSION AND SERVICE PARAMETERS

The transmission parameters described above can now be classified according to their related effect on the quality of a telecommunication service, see Figure 3.7. Many of the type A factors (cf. p. 17) mainly influence the one–way voice transmission quality. Amongst these are e.g. the loudness of the transmitted signal (SLR and RLR), different types of noise (ambient noise, continuous circuit noise and impulsive noise), distortions due to the frequency shape of the transmission channel and codecs, as well as transmission errors (bit errors, packet or frame loss) and listener echo. However, it can be argued that severe problems in the one–way transmission do influence the conversational behavior as well. A too loud or too quiet connection, e.g., will force the talker to adapt the speech level. This has an influence on the ease of communication. Parts of the discourse which are lost because of transmission errors result in a question of reconfirmation ("Did you say that...?") and influence the flow of the dialog. Therefore, all such type A factors also exercise an influence on the communication efficiency. The latter is directly influenced by factors of type B and C which occur only for talking and listening, like talker echo, talker sidetone, absolute delay or longer interruptions in the transmission. Many of the new types of degradations (see Table B.1) are of type A, and first exercise an influence on the one–way voice transmission quality, but in a second step also on the communication efficiency. Talker echo, talker sidetone and absolute delay are closely related to the conversation effectiveness, rather than to voice transmission quality.

Other factors which have not been discussed here in detail relate to the service comfort and the costs. The comfort is e.g. influenced by the type of terminal including its ergonomic design, as well as by factors like availability, reliability, and set–up and response times. The advantage of access from a mobile terminal is classified as an advantage in service comfort. Costs include the investment costs (e.g. for the terminal equipment), operational costs for each connection, as well as the account conditions, see Section 7 of Chapter 4.

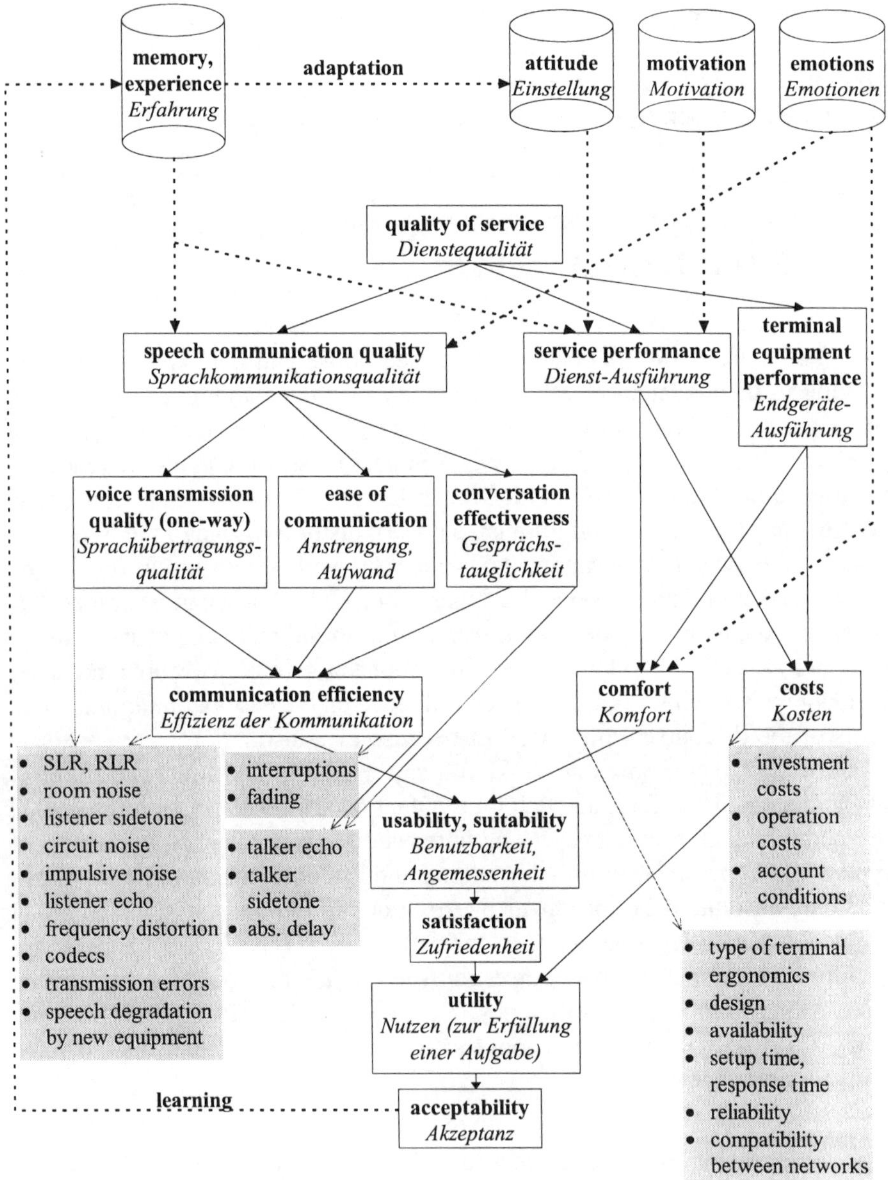

Figure 3.7. Classification of transmission parameters in the QoS schematic. German terms are given in italics.

5. SUMMARY

The quality of telephone speech is fundamentally influenced by the characteristics of the transmission system. Telephone transmission includes a large variety of different types of equipment. Thus, it is not reasonable to analyze every piece of equipment (terminal, switching and transmission components) in full detail.

The approach made in this chapter is to discuss perceptively relevant factors for the quality of a telephone service by reviewing some underlying psychoacoustic models. The perceptive factors were chosen in order to describe signal–related characteristics. They are associated with traditional systems and include loudness, articulation, the effects due to bandwidth restriction and frequency distortion, the perception of sidetone and echo, the perception of various noises and delay. A psychoacoustic loudness model according to Fletcher and Munson was presented which forms the basis for the principal description of attenuation and frequency distortion used in telephony, the so–called loudness ratings. The fundamental assumption of the model is that contributions to loudness can be calculated from the according sensation levels. The overall loudness is the integral of all contributors over the relevant frequency range. A similar model (based on Collard's findings) was given for articulation. The same principle will also be used later for predicting speech communication quality, cf. Section 1 of Chapter 5. For the other perceptive factors, no such psychoacoustic models have been set up so far. Especially new types of equipment cause several degradations to speech quality which are not encountered in traditional networks. These new degradations have not been analyzed in full detail. However, a classification is proposed which will allow to cover these effects in network planning, see Appendix B.

Common to all transmission paths involved in a telephone connection (the principal transmission path, the sidetone path and the echo path) is that their associated effects are mainly expressed in terms of loudness ratings. Scalar frequency weighted factors can also be calculated for the different (circuit and ambient) noise components. The description of the transmission system by means of a set of scalar parameters thus corresponds to the perceptive factors. This parameter set gives a useful though not complete view of a general telephone configuration. It will later be taken as an input to models which try to predict speech communication quality from the transmission characteristics of the telephone network.

In order to investigate the quality of telephone speech it is convenient to set up a simulation model rendering the perceptive factors identified so far experiencable to human talkers and listeners. The simulation was planned and implemented by the author, mainly as a tool for carrying out auditory quality assessment experiments. All relevant characteristics of the transmission system can be manipulated in a defined way. In addition to the traditional equipment, also various low–bitrate codecs are simulated. As an input to the simulation model frequency characteristics as well as scalar values (loudness ratings, various noise levels) are permitted. It was implemented on a DSP platform and integrated in a complete test environment. The real–time capability permits listening–only and conversational experiments to be carried out using identical circuit characteristics, and thus allows a direct comparison of the results obtained with different test methodologies. The test methodologies are further discussed in the following chapter.

The transmission parameters were classified regarding their effect on different quality of service related components. For this purpose the QoS schematic established in Chapter 2 was used. It turns out that, due to their nature, the transmission parameters mainly exercise an influence on the one–way voice transmission quality or (for echo, sidetone, delay and factors deteriorating the flow of the conversation) on conversation effectiveness and communication efficiency. Other parameters relating to the service affect comfort and costs, and thus usability or utility. The classification will later allow appropriate quality assessment methods to be chosen, in order to investigate their related effects. It also provides insight into the dimensions of the quality of service which are estimated by prediction models.

Chapter 4

QUALITY ASSESSMENT IN TELECOMMUNICATIONS

If the perceptive effects of the factors discussed so far are to be qualified or quantified, auditory assessment methods are needed. Besides the test object, the choice of an adequate test method is the principle influencing factor on what information on quality can be obtained as the result of an assessment. It is therefore necessary to discuss in more detail the principle quality assessment methods and methodologies used in telephony, and to point out which consequences result for the outcome in terms of quality dimensions. In Sections 1 to 8 various assessment methods are analyzed regarding the perceptive dimensions they refer to. The analysis includes the influences resulting from the choice of test subjects, the test procedure, the scaling method, the test modality, etc. It turns out that there is no optimal test method for all or for general purposes, but that an appropriate method has to be chosen according to the assessment task. With respect to the quality dimension which is assessed, the test methods can be classified in the QoS schematic defined in Chapter 2, see Section 10. In order to get valid estimates of integral speech communication quality in telephone situations, an economical conversation test methodology is developed, see Section 9. This methodology is used in the following in order to investigate the relationship between factors governing the quality of a service (Chapter 6) and the quality of prediction models (Chapter 7).

There are several ways of categorizing assessment methods, e.g. regarding the modality of the test (listening–only or conversational), the rating procedure (single stimulus or comparison, adjustment or constant method) or even the test environment (laboratory test vs. surveys or field tests). In this chapter, a categorization is preferred that allows statements on the quality entity which is assessed by a specific type of assessment method to be made, as it regards the schematic presented before. For this purpose, it is worth citing a classification of assessment methods from Richards (1973) first:

- Class I: comprises those methods that can be applied for assessing the widest possible range of types of speech links, degraded in any manner. Example: Conversation tests (CTs).

- Class II: comprises methods that are less general than those in class I, but are very commonly applied because they are correspondingly less laborious and time consuming to conduct. Most but not necessarily all class II methods require the use of subjects, but a special team trained for the work will sometimes suffice. Example: Listening–only tests (LOTs) with sentence material, intelligibility tests, threshold determination, etc.

- Class III: consists of methods which seek to describe relevant fundamental characteristics of particular speech paths. Example: Instrumental measurements of sensitivity/frequency responses, noise level measurements, etc.

Strictly speaking, class III methods do not refer to 'quality' assessment, because quality does not exist independently of a subject and thus cannot be assessed without performing tests or investigations with subjects. Without questioning the usefulness of instrumental quality predictors, models can only predict what they were trained or designed to, namely the results of subjective tests of class I or II. On the other hand, the classification given above mainly refers to tests under controlled laboratory conditions. General quality assessment methods additionally permit usability, utility or acceptability to be evaluated. Some information on the methodology for such evaluations is therefore worth noting, and this will be done in Sections 5, 6 and 7.

Quality assessment makes wide use of psychometric methods, i.e. methods which aim at measuring perceptual objects. A general categorization of psychometric methods (= psychophysical and psychological measuring techniques) can be made regarding

- the scaling method and the resulting level of the measurement scale:

  - Magnitude estimation, ratio estimation, magnitude production, ratio production (constant or adjustable stimuli). The resulting scale can have ratio level.

  - Category estimation, category production, preference judgment, similarity judgment. The resulting scale level is either ordinal or interval, depending on the assumptions made to the distribution of stimulus responses and the correlation between responses to different stimuli (Guilford, 1954; Sixtl, 1967; Thurstone, 1927).

  - Performance tests, e.g. articulation tests. The resulting scale level is nominal.

- the presentation method: Method of adjustment (by subject, by the experimenter), constant or interrogative methods (single stimulus, paired comparison).
- the 'modality': Listening–only tests (LOTs), conversation tests (CTs), talking and listening tests.

Part of this categorization can be seen later in this chapter in the description of quality assessment methods used in telephony. However, it is not the aim of the author to provide a detailed description of general psychometric considerations. Reference should be made to the psychophysical literature (Borg, 1994; Borg and Staufenbiel, 1993).

It is very difficult to give general guidance on which assessment method or methodology is adequate for a specific quality assessment problem. Nevertheless, a general classification attempt of assessment methods regarding their applicability to specific tasks will be made here. The classification reflects experience gained after using these methods for a number of years. The main conclusion drawn is that there is no general 'best' assessment method or methodology independent of the test task. The optimum method has to be found for each specific task by an experienced test designer. This includes the choice of test stimuli, subject group, number of objects to be tested, and other environmental factors. Also, it is not recommended to use a standardized test method or scale if there is a more appropriate one available. On the other hand, standardized methods and scales reflect some of the experiences gained and compromises made between applicability and quality of test results. The use of such methods may be more appropriate if test results obtained on different occasions have to be compared, and if there is no time to find and verify a better method. The price, however, may be a sub–optimal quality of the results, in terms of the information content.

To obtain highly valid integral quality judgments, conversation methods are generally preferable for telephone quality assessment. If a detailed analysis of quality dimensions or test economy require the use of listening–only tests (LOTs), paired comparison (PC) is generally recommended for stimuli that are almost equal, because the differential sensitivity of the ear is much higher than the absolute sensitivity. However, only an ordinal or interval judgment (cf. Section 8) can be obtained, and only a few stimuli can be tested within one test session. Absolute category rating (ACR) is recommended for more varied test stimuli, and for obtaining more 'absolute' judgments (cf. Section 3.1). If valid results for high quality conditions are required, the comparison techniques (degradation category rating, DCR, or comparison category rating, CCR) may be more appropriate in differentiating between conditions. As a rule of thumb, DCR is said to be more sensitive in the range $Q \geq 20$ dB (Rosenberger, 1988).

The differential sensitivity of the ear can be further increased by forced choice comparison (no 'equal' judgment). An alternative method to ACR and PC is the paired rating method (PR), recommended e.g. for listening tests on loudspeakers (IEC Technical Report 60268-13, 1998). A description of PC and PR methods is given in Section 3.2. If more analytical information than just an integral quality judgment is required, multidimensional analysis can yield highly informative test results. It is also possible to ask test subjects more analytical questions in a listening–only, sometimes even in a conversation experiment. However, in this case it is often more appropriate to use trained test subjects.

Laboratory tests always lack some naturalness, in the sense that the assessing subject is in an artificial situation when reporting his/her opinion on the test items. Especially in telephony, the motivation for simulated test calls is different from real life. These differences influence the test results, especially when more user–directed factors are investigated (utility, acceptability, cost–related factors). For those cases, it is more appropriate to perform user surveys or to use usability measures, as presented in Sections 5 and 6.

1. CHOICE OF TEST SUBJECTS

The selection of listening or conversing subjects should be guided by the test purpose. In order to obtain valid integral quality judgments, the subject group should have a relation to the population which will use the system under test. Therefore, the subject group for general telephone quality assessment tests should represent the average telephone user population as accurately as possible.

In general, untrained subjects will yield a more valid judgment of 'absolute' quality levels in the sense that they are not influenced by knowledge about circuits and the selection of conditions used in the test. On the other hand, experts or trained subjects can produce more analytical results. It has been observed that results of preference judgments are affected by learning and adaptation effects (IEEE, 1969, p. 230). For intelligibility tests, it is often preferred to have subjects from whom stable results can be obtained in consecutive test sessions, but also in this case the choice of test subjects depends on the aim of the experiment.

The use of fewer, trained experts as test subjects allows the time and costs needed to carry out experiments on telephone circuits, both in the listening–only and conversation mode, to be reduced. These 'informal' expert tests are very helpful in order to get a quicker indication of conversational quality. They are used e.g. to choose test conditions from a larger number of potential conditions for a consecutive formal test, to get an initial indication of the quality for a small number of test conditions, or to obtain more analytical and diagnostic information in order to investigate a specific problem (Arden and Reeves, 1996). Especially for the latter purpose expert assessment can yield more diagnostic information than could be obtained from naïve test subjects. However, it should always be kept in mind that judgments from experts may be highly biased. They are therefore not comparable to those obtained from naïve subjects (which ultimately represent to a larger extent the normal telephone user group). Informal tests tolerate the abandonment of certain principles necessary for a formal test in order to obtain highly reliable and valid results. The choice of which principle may be abandoned in order to get rapid though valid results is the responsibility of the test designer. The high demand for these informal tests with experienced subjects is reflected in the inclusion of such tests in new ITU–T Recommendations regarding test procedures e.g. for echo cancellers (ITU–T Draft Rec. P.831, 1998).

2. ARTICULATION AND INTELLIGIBILITY TESTS

The sensible use of articulation and intelligibility tests is limited to connections with rather poor transmission performance. Therefore they are very rarely used for assessing modern telecommunication networks. In special cases the methods can provide one characteristic of the speech link which is not always covered by more general integral quality tests. For example, in standard LOTs, the listener does not have to make an effort to understand the meaning of what is said. It should be noted that articulation and intelligibility tests yield information that differs greatly from the one obtained on the listening–effort scale in absolute category rating tests (see Section 3.1). The reason is that it is a different task for the listeners to give an impression of perceived listening–effort, or to write down what they have heard. Like loudness, results of articulation measurements relate only to type A factors which cause difficulty in listening. More information on articulation and intelligibility test methodologies can be found in Appendix C.

3. LISTENING–ONLY AND CONVERSATION OPINION TESTS

The only definitive method for testing bidirectional systems is conversation methods, such as conversation tests or user surveys (ITU–T Handbook on Telephonometry, 1992; Richards, 1973). Listening–only tests may be used to assess specific factors of type A (attenuation, noise, quantizing distortion), but care should be taken that a satisfactory correlation exists between the LOT and CT results for that specific impairment (cf. Section 3 of Chapter 6). Attention must be paid when investigating time-varying impairments: at least one complete time cycle must be present in the test stimulus. The assessment of time–varying impairments is a subject for further study.

Talking–only tests show a poor correlation to CT results for talker echo and sidetone, because the talking–only situation is an unrealistic one. Therefore, this method is generally not recommended by the ITU–T (ITU–T Handbook on Telephonometry, 1992, p. 110). However, some interesting results from a comparative study of different test methods show that talking and listening tests may provide analytical information, cf. Section 3.4.

In the following sections, different types of listening–only and conversation tests will be discussed in more detail. These tests investigate either integral communication quality or single aspects contributing to speech communication quality. More general approaches aiming at investigating usability or utility will be discussed in Sections 5 to 7. Details on the theory of scaling and on conversation tests scenarios follow in Sections 8 and 9.

3.1 LISTENING–ONLY TESTS USING ABSOLUTE CATEGORY RATING

LOTs using sentence material are the most common test method for investigating integral quality. In absolute category rating (ACR), each stimulus is rated individually on a category scale. The category labels are chosen in order to provide some kind of 'absolute' level information. The validity of test results regarding an 'absolute' quality horizon will be discussed later in Section 1 of Chapter 6. One fact is, however, that LOT results are strongly influenced by the choice of conditions in the test material. Because the listening–only test situation is a very artificial one, and because the subjects in a LOT mainly focus on the form of the speech sound and not on the contents, the influence of the stimulus condition range (and thus the quality range) is stronger than in conversation tests. To avoid this drawback, a sufficiently large range from worst to best potential quality events should be included in the test stimuli. An initial reference can be established before the test starts, by playing some dummy conditions beforehand. However, it should be noted that this has a great influence on the test subject's use of the scale. Especially, the 'absolute' character suggested by the category labels is partly destroyed, and the test subjects will try to match the range of (potential) test conditions to the scale range available.

The test set–up and test run are well described in the relevant ITU–T literature (ITU–T Rec. P.800, 1996; ITU–T Handbook on Telephonometry, 1992). Various scales are recommended for use, which focus on different dimensions of quality. Graphical presentations of three different scales are given in Figures 4.1 to 4.3. All scales used here are pre–annotated with numbers. The advantage in contrast to post–annotation (after the experiment) is that the numbers support the interval character of the scale, suggesting that the categories are equally spaced and of equal width. Advantages and drawbacks of different category scales are further discussed in Section 8. The mean over the ratings obtained on a listening–quality scale (Figure 4.1) is denoted by Y_{LQ}, and is generally called the mean opinion score, MOS. However, the abbreviation MOS often also refers to conversation tests. Therefore, if ambiguities occur, Y_{LQ} or LOT MOS will be used to give greater precision. The mean rating on a listening–effort scale (Figure 4.2) is denoted by Y_{LE}. The term listening–effort applies to the perceived speech signal and not to the system as such. Y_{LE} is thought to measure some kind of 'intelligibility', like an 'intelligibility question' in a polarity profile, or mental effort. Mean preferred loudness ratings are denoted by Y_{LP}, cf. Figure 4.3. The entity measured on this scale is some (5...7) dB quieter than 'optimum loudness' for which the highest Y_{LQ} score can be observed. Both 'preferred' and 'optimum' loudness depend on the room noise, the bandwidth and other factors.

Quality of the speech:

excellent	good	fair	poor	bad
5	4	3	2	1

Figure 4.1. Listening–quality scale according to ITU–T Rec. P.800 (1996).

Effort required to understand the meaning of sentences:

complete relaxation possible, no effort required	attention necessary; no appreciable effort required	moderate effort required	considerable effort required	no meaning understood with any feasible effort
5	4	3	2	1

Figure 4.2. Listening–effort scale according to ITU–T Rec. P.800 (1996).

Loudness preference

much louder than preferred	louder than preferred	preferred	quieter than preferred	much quieter than preferred
5	4	3	2	1

Figure 4.3. Loudness preference scale according to ITU–T Rec. P.800 (1996).

Apart from the quality range, the type of test stimuli is another factor influencing listening–only ACR test results. The ITU–T Handbook on Telephonometry (1992) recommends the use of groups of 2–5 isolated, short sentences. In this way, test subjects will not concentrate on the contents after they have listened to a few test stimuli. The attention focuses only on the form of the signal, which is unrealistic for normal telephone use. To reduce the effects of this drawback, stimuli of tests carried out in this work consisted of short text paragraphs which briefly describe a case or an event. Extracts from EUROM.1 text passages were used for this purpose (Gibbon, 1992). In a listening–only test, it is not the circuit characteristic which is assessed, but the perception of a speech sample. The characteristics of the stimulus will thus depend to a large extent on the speaker. Bennett and Lincoln (1985) show that there is a large spread between the judgments on test stimuli from different speakers degraded by the same circuit. This spread is unlikely to be balanced out by using only 4–6 different speakers in a listening test. Moreover, a strong speaker by circuit interaction exists. I.e., the difference between speakers affects perceived quality differently for e.g. some LPC codecs than for e.g. multiplicative noise. Especially codecs will show that speaker by circuit interaction in an unpredictable way. If weaker speaker dependence is to be expected, a minimum of two male and two female speakers may suffice.

3.2 LISTENING–ONLY TESTS USING PAIRED COMPARISON TECHNIQUES

Whereas ACR methods tend to show low sensitivity when approximately equivalent conditions are rated, degradation category rating (DCR) and comparison category rating (CCR) offer a higher sensitivity. These techniques have been adopted from audio testing of high quality circuits. Both are paired comparison ratings where stimuli are presented in pairs (AB) or repeated pairs (ABAB), pronounced by the same speaker. In DCR, an unprocessed high quality reference is presented as sample A. Subjects have to rate to what extent the second processed sample B is degraded in comparison to the first. In CCR, processed and unprocessed samples are chosen at random, and subjects rate the quality of the second sample (processed or unprocessed) in relation to the first. The scales which are recommended by ITU–T are given in Figures 4.4 and 4.5.

Degradation

degradation is inaudible	degradation is audible but not annoying	degradation is slightly annoying	degradation is annoying	degradation is very annoying
5	4	3	2	1

Figure 4.4. Degradation category rating (DCR) scale according to ITU–T Rec. P.800 (1996).

The quality of the second compared to the quality of the first is:

much better	better	slightly better	about the same	slightly worse	worse	much worse
3	2	1	0	-1	-2	-3

Figure 4.5. Comparison category rating (CCR) scale according to ITU–T Rec. P.800 (1996).

The higher sensitivity of paired comparison ratings is partly due to the more linear rating scale (DCR) in comparison to the MOS scale (ACR). A second effect is that in DCR processed and unprocessed conditions of the same speech sample (text and voice) are rated in relation to each other. Quality differences between voices are therefore balanced out (Gleiss, 1987b).

The isopreference method is a further method for determining preference between paired speech samples, and thus indirectly the speech quality. It is based on multiple forced–choice comparisons of speech samples at several signal–to–noise ratios, and the subsequent establishment of contours in the speech level — noise level plane. For the established contours, a metrics can be defined in terms of a "Transmission Preference Level" or "Transmission Preference Units". The method was established by Munson and Karlin (1962), and a simplified version was described by Rothauser et al. (1968). It will not be discussed in detail here, because of one major drawback, namely that it uses the speech level — noise level ratio as an explicit reference. In the investigations of Munson and Karlin (1962) real live speech at 1 m airpath is used as a reference, which is compared to speech transmitted over electrical circuits. This real live speech, as well as speech degraded by additive or multiplicative noise is not an appropriate reference for many kinds of degradation occurring in telecommunication networks. The same drawback exists for the relative preference method as described by IEEE (1969). Isopreference tests can also be set up in such a way that the test subject can adjust the strength of a reference degradation in order to match the quality or impairment of a test speech signal. This method has been used to test the influence of different references on a quality rank order in Section 2 of Chapter 6.

The paired rating (PR) method can be seen as a combination between ACR and PC rating techniques. A description is given in IEC Technical Report 60268-13 (1998). The listeners are asked to rate the attributes (one or more) of each of the stimuli individually on a scale (no preference judgment!). The subjects can switch between the two stimuli presented in a pair according to their desires. Experience shows that both PC and PR methods are often preferred to ACR by non–trained listeners. The use of PR in telephone quality assessment, however, has not been reported on so far.

3.3 MULTIDIMENSIONAL ANALYSIS

If more analytical information than just an integral quality judgment is desired, the test design should capture different aspects or dimensions of quality. Generally speaking, two contrary approaches are commonly used in a listening–only procedure. In multidimensional scaling the different quality dimensions are captured in the test material. During the test subjects are asked to rate the similarity, dissimilarity or even their preference for test stimuli presented in pairs. A later factor analysis allows the test conditions to be described in a multidimensional space. The dimensionality of the space is largely determined by the test material. However, very often test subjects base their similarity or preference judgments on approximately 2–3 internal dimensions, which are the principal components of the factor analysis capturing 85–90 % of the inherent variance. It is the experimenter's task to associate the dimensions with attributes allowing different aspects of quality to be described, as no information on the interpretation of the judgments is given by the subjects.

A complementary approach is to define a set of scales which capture most of the differences in the speech material. The scales should be meaningfully labeled, and the interpretation of the labels should be made perfectly clear to the test subjects. In the test, ratings on each of the scales are obtained for each test signal and subject. A very commonly used scale type is a continuous bipolar scale with antonyme labels on each extremity. This scaling technique, called the semantic differential technique, was developed by Osgood et al. (1957) to analyze meaning in language, but is now used in many fields of psychological research. A detailed description can be found e.g. in Schäfer (1983). The judgments on each scale can be documented in the form of a polarity profile, i.e. the mean ratings on the single scales are listed below each other and connected, resulting in a profile specific to the test condition and to the set of scales used. Interpreting the results is easier than in multidimensional scaling as the labels help to describe different dimensions. However, it cannot be guaranteed that all the dimensions in the test material are captured by the set of scales used, and that all the subjects use the same interpretation for the scale attributes (labels). Many of the attributes are correlated, a fact which makes the test procedure less economical.

Multidimensional assessment of speech signals sometimes, but not necessarily, makes use of a parametric evaluation approach. In the more common 'isometric' approach opinions on quality are directly solicited from the listening subjects. It is implicitly assumed that the sample of listeners is representative of the population of interest, in that the distribution of influencing factors on the quality judgment is well represented. The 'parametric' approach used in multidimensional assessment only requires the sample of listening subjects to have a representative capacity to discriminate a set of perceptive characteristics in the speech samples, but is not necessarily representative when making quality judgments. The subjects are not asked for their preference for these characteristics. Instead, it is assumed that a correlation exists between the perceptive characteristics and the quality judgment of the population of interest. The quality estimates are then calculated from the perceptive characteristics of the speech signal. The disadvantage of such a method is that the relation between the perceptive characteristics and the quality judgment has to be established first, depending on the characteristics of the speech samples to be assessed. Especially, this relation will be highly dependent on the application for which the assessment has to be made.

It is well known that scores obtained using the 'isometric' approach generally show large inter–individual variations. This variation results from both random errors in judgment and differences in the importance of perceptual characteristics on the quality judgment. The parametric approach is thought to be more accurate because the latter differences can be avoided. However, the validity of the assessment is more critical.

A methodology which makes use of the 'parametric' approach and tries to combine it with 'isometric' evaluation is the Diagnostic Acceptability Measure (DAM). It was introduced and developed by Voiers (1977), and a detailed description can be found in Quackenbush et al. (1988). In principle, it is a refined version of the polarity profile method, using highly trained expert subjects. Only candidate listeners who show a moderate to high correlation to the listening crew's historical average over a large and diverse set of training utterances are used´. Their ratings are calibrated onto the historical average ratings. Listeners rate selected utterances on 20 response scales, ten of which relate to the speech signal, seven to the background, and three to overall quality. Ratings on the 17 scales related to signal and background characteristics are then combined to form scores on ten reporting scales. Two additional reporting scales evaluate total signal quality and total background quality. To these twelve parametric reporting scales four isometric or metametric reporting scales are added, which give an indication of "intelligibility", "pleasantness", "acceptability", and the main result, the so–called "Composite Acceptability (CA)". This latter value is an effective summary of the ratings on all the other scales. It should be mentioned that the term "acceptability" as used by Voiers and Quackenbush is not comparable to the definition used here. In contrast, their definition does not cover any aspect of an actual system or service, but is a more general approach to establish an accurate assessment method independent of the application situation. As quality has been defined as the extent to which the users expectations and desires are catered for, the validity of the final CA score may be restricted. The DAM has been developed "including virtually every present day digital voice system" (Voiers, 1977), but may not be applicable to a specific system or service, or even to a type of degradation it has not been designed for (e.g.

time–variant degradations). On the other hand, the parametric reporting scales can provide additional analytical information which will be of great use for interpreting the effect of certain systems on transmitted speech.

3.4 TALKING AND LISTENING TESTS

Talking–only methods are sometimes used to assess the detectability or objectionability of talker echo or sidetone. In the ITU–T Handbook on Telephonometry (1992) it is stated that the results of these tests do not correlate well with those obtained in conversation, because the talking–only situation is a completely different (and rare) one, thus leading to a different subject behavior. A similar effect was found in the detectability testing on absolute delays, where trained subjects were able to detect delays much earlier (i.e. at lower delay times) than untrained subjects in actual conversation situations. The use of talking–only methods therefore does not give a realistic indication of the quality of the factors of interest in a conversation situation.

In an investigation performed by Deutsche Telekom and Head acoustics GmbH, the results of a conversation test, a listening–only test and a specific double–talk test were compared to assess both overall quality and several distinct quality parameters in a connection involving hands–free terminals (ITU–T Delayed Contribution D.023, 1997). In each test trained as well as untrained subjects took part. The listening–only test set–up was specifically designed in order to examine the hands–free terminals, using a head and torso simulator (HATS) with an artificial mouth. The double–talk test consisted in a conversation where one test subject was reading a text and the other subject interrupted for a short period of time. Both subjects experience the double–talk section in a different way (subject no. 1 while talking and subject no. 2 while listening), so different questions can be asked to the test subjects. The test procedures are described in more detail in ITU–T Contribution COM 12–5 (1997). It turned out that conversational and double–talk tests lead to different results for the trained and untrained test groups. The differences depended on the parameters to be assessed. For the parameters which were investigated in all the tests (e.g. overall quality in terms of MOS), double–talk and conversation test results correlated highly with listening–only test results, but only for the trained subjects' group. For the untrained subjects these ratings differed significantly. Apparently, the choice of the test group plays a major role in assessments using double–talk tests. There is no general high correlation to the results of conversation tests. On the other hand, methods like the double–talk test can be used in order to investigate specific characteristics e.g. of some terminal equipment. In this respect they may provide additional information which cannot be obtained from general purpose conversation tests.

3.5 CONVERSATION TESTS

With few exceptions, telecommunication systems are designed for the bi–directional exchange of information, and this is what they are mostly used for. Quality assessment of such systems or services should therefore reflect the bi–directional exchange, and the change in the talker and listener role. This can only be performed in a natural conversation situation. Thus, conversation tests are necessary when overall quality judgments of high validity and reflecting an 'absolute' quality horizon are required. A natural test situation will stimulate the natural behavior of the conversing subjects. Under perfect circumstances, test subjects should have a natural conversation including a natural beginning and ending, and a natural task to stimulate the two–way conversation.

The naturalness of the situation has an influence on the assessment task. Subjects will not primarily concentrate on the form of the speech signals, but on what is said, i.e. the contents. The focus of attention implies that not a great deal of diagnostic information can be expected from the test subjects. ITU–T Rec. P.800 (1996) proposes two scales for conversation tests: a 5–point conversation quality scale and a binary difficulty scale. Mean subjects' ratings on these scales are denoted Y_C and %Diff. Graphic representations are given in Figures 4.6 and 4.7.

The choice of test conditions has a minor impact on the quality assessment results compared to LOTs. The test situation suggests to use the personal history in using telephone systems as a reference for the opinion ratings, and only to a lesser degree the other conditions included in the test. Nevertheless, the range of test conditions should contain quality likely to be considered from "excellent" down to "bad". The use of dummy conditions before the test starts helps to familiarize the subjects with the test procedure, but will certainly exercise an influence on the way they use the scale. In a conversation test, a sufficient number of voices is used in order to minimize effects like talker–speaker interaction, range of speech levels, etc. In this respect, conversation tests are much more equilibrated than LOTs.

Opinion of the connection you have just been using:

excellent	good	fair	poor	bad

Figure 4.6. Conversation quality scale according to ITU–T Rec. P.800 (1996).

The experimental design should help to minimize all the potential biases: order within the experiment, conversation scenario, room, conversation partner, etc. Graeco–Latin squares (GLSs), a special method for designing incomplete

Did you or your partner have any difficulty in talking or hearing over the connection?

☐ Yes ☐ No

Figure 4.7. Difficulty scale according to ITU–T Rec. P.800 (1996).

test plans, are adequate for that aim. In GLSs, not all the potential combinations of biases are equilibrated. However, a lot of test subjects and sessions are saved in comparison to complete test plans. An example of a GLS test design is given in Appendix D. A simplified method of a test plan is to establish two or more randomizations of circuit conditions, multiplied by two (or more) randomizations of conversation scenarios. This prevents biases, while reducing the time necessary for the experimental set–up (preparation of test sheets in different orders, preparation and verification of test condition lists, etc.).

The major drawback of full conversation tests is that they require a great deal of time and money expense to be performed. These expenses are only partly offset by the fact that no test signals have to be prepared (only instructions, test lists, etc.). In order to avoid these major inconveniences (which sometimes render the use of CTs impractical), it is possible to conduct

- 'simplified conversation tests' without dialing and ringing. Switching between connections allows a conversational task to be spread over more than one connection.
- 'short conversation tests', limiting the time needed for a conversation by means of the conversation task, while maintaining the full conversation character. Conversation tasks for such tests were developed in this work, cf. Section 9.
- 'informal' experts tests (cf. Section 1).

Especially the 'short conversation tests' (SCTs) greatly reduce the major inconvenience of ordinary conversational tests. Thus, more valid quality assessment results can be obtained. The conversational task will be discussed in detail in Section 9. Most of the tests for investigating speech communication quality in the present work were performed in this way.

4. PERFORMANCE TESTS

A user's aim when establishing a telephone connection is to transmit or exchange information with his/her communication partner. Generally speaking, the telephone connection serves a specific purpose or task, which normally differs for every individual call. In laboratory testing on telephone connections, the efficiency in performing a task over a telephone connection can be taken as an approximate index of the speech communication quality. The task is then specifically designed for such a laboratory test. This method is called performance test.

Some examples of performance tests are given in the ITU–T Handbook on Telephonometry (1992). One simple method is to count the number of check questions of a recorded conversation, by choosing only those check questions which probably result from bad transmission quality. For highly developed tasks it is also possible to count the number of errors which occur in resolving a task, or to measure the time needed to resolve a task. The main disadvantage of such methods is that they are largely user dependent and task dependent. Many different factors have an influence on the performance measures which do not even originate from the transmission. E.g., it is very difficult, even for trained experimenters, to decide whether check questions in a conversation result from bad transmission quality, or whether they are due to the test subject being distracted, ill–formed articulation of the talker, errors in reasoning, etc. An additional disadvantage is that a measurable degradation of task performance only occurs in very bad connections. For these reasons, performance tests are only rarely used in assessing modern telephone network quality.

5. USER SURVEYS

When the artificial laboratory situation is expected to dominate the quality evaluation, user surveys can be performed to assess the customer opinion on telecommunication systems. Generally, user surveys can be expected to yield more 'absolute' results in the sense of the naturalness of the test situation. However, practical implications allow only in extremely rare cases the transmission characteristics of the communication line to be controlled. Therefore, only mean results over an 'average' system can mostly be obtained.

ITU–T Rec. E.125 (1998) recommends two versions of questionnaires for investigating the quality of service of international calls, one for national subscribers and another for foreign visitors. The questions were designed to determine basic sources of user difficulty which may arise when making a call, mainly during dialing and call set–up, but also related to the transmission quality. Transmission quality related questions evaluate the overall quality of a specific connection as well as the difficulties in communication. As foreign visitors are expected to stay in the country only for a limited period of time, the questions for this group relate mainly to their last call, thus to their short–term experience. The questionnaire for national subscribers provides information regarding a more general long–term experience with the telephone service. It is recommended that these questionnaires be used on a cross–section of subjects (residential as well as business subscribers), and that small samples of one class in the whole subject group should be avoided. The questions are designed to be presented aurally (either face–to–face or over the telephone) by experienced interviewers, who should understand technical terms in order to realistically classify subjects' responses. Examples of the survey questions can be found in ITU–T Rec. E.125 (1998); ITU–T Rec. P.82 (1989).

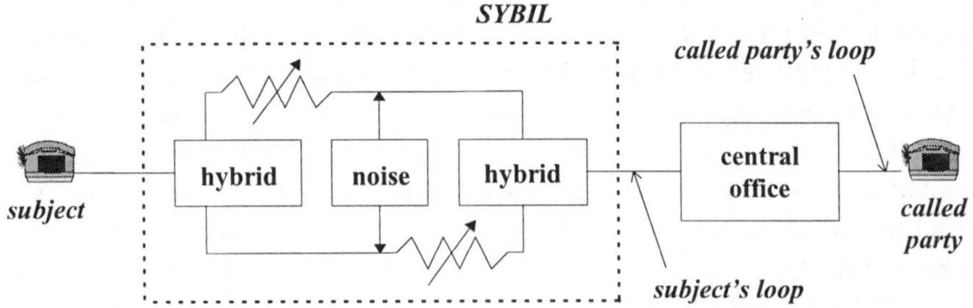

Figure 4.8. Diagram of a telephone connection with SYBIL (ITU–T Suppl. 5 to P–Series Rec., 1985).

An approach to overcome the limitations of user surveys regarding the lack of control over circuit conditions was the SYBIL system installed at the Holmdel Bell Laboratories in the 1970s. The system allowed an experimenter to control transmission parameters during normal business calls from cooperating Bell Labs employees. A schematic of the set–up is given in Figure 4.8, which is a reproduction of Figure 1 of ITU–T Suppl. 5 to P–Series Rec. (1985). Cooperating subjects received certain transmission characteristics which were inserted automatically into some of their general business connections (however, subjects could choose whether they wanted to make a normal call or a test call). If the transmission characteristics were unacceptable, the subject could switch to a normal call by dialing a special digit. After a test call had been completed, the test subject received a ring back and then could make his/her judgment by dialing a single digit indicating the rating on a 5–point ACR scale (MOS scale). This method guaranteed that the test subjects had a natural motivation for making the call, and that they judged the quality under the impression of that specific call. The subjects' choice between performing a test call or a normal call resulted in subjects being willing to contribute to the assessment, a fact which is important for obtaining reliable ratings. However, one drawback remains, namely that the situation mainly refers to business calls. A similar experimental set–up with residential subscribers would certainly be very costly, both in terms of the equipment as well as for the organization of the experiment.

Without additional costs, test subjects participating in laboratory (conversation or listening–only) experiments can give information about the general opinion and behavior when using the standard telephone system. These questionnaires are no user surveys in a narrow sense, but they provide an image of general user requirements and opinion. Questions can be divided into those relating to the general behavior when using the telephone system, and others relating to special telecommunication services (mobile phones, telephony over IP) or to the price. Examples of questions used within different tests of Chapters 6 and 7 are given in Appendix F. It should be noted that all the questionnaires have the drawback that a pre–selection of potential responses is made when designing the questionnaire. Other methods can overcome this inconvenience, e.g. the "Intermitting Loud Thinking" (Schulte-Fortkamp, 1994), see Section 6. Regarding the validity of price–related questions reference should be made to Section 7.

6. USABILITY EVALUATION

The usability of a telecommunication service is related to the communication system as well as to the service. The system is mainly characterized by its speech communication quality, but also by the ergonomic design of the user terminal, etc. The service is used to fulfill a specific task or objective within a specific environment. The usability evaluation will therefore depend on the task, user and environment, and results of usability evaluation should always be seen in this context (Gleiss, 1992).

As a global categorization, measures of usability can be of two kinds. They can focus on the task performance, i.e. how well a user can fulfill a specific task, or they can focus on the user's opinion, i.e. how well he or she likes to use the system (ETSI Technical Report ETR 095, 1993). These two factors can be considered to be independent, in the sense that a system can achieve high task performances, but users may have a bad opinion of using the system. They are related, on the other hand, by sharing the same system. Usability evaluation should therefore focus on both usability dimensions.

In order to develop highly usable products, a first step is to review design guidelines in the form of a 'checklist'. Such guidelines exist for a set of terminal equipment used in telephony, cf. ETSI Technical Report ETR 051 (1992); ETSI Technical Report ETR 147 (1994), as well as for user interfaces with man–machine dialog (Bernsen et al., 1996). The application of these rules, however, will not always guarantee a highly usable product or service, especially when new systems or services are to be developed. Then, usability testing is required. The most common methods for testing, as described by ETSI (Gleiss, 1992), give an overview of the steps which have to be performed when developing a highly usable product. They are divided into the "objective, non–intrusive methods" like secondary sources, logging and observation on the one hand, and the "subjective, intrusive methods" on the other hand (questionnaire, interview, self–descriptive). The following methods are the most commonly used ones (ETSI Technical Report ETR 095, 1993):

- Secondary sources: Already existing data which was neither collected by the usability assessor nor for the purpose of the evaluation is used to improve the usability of a product. Examples: Problems with existing systems or services, statistics on user behavior, problem reports by users, etc.
- Logging: A recording device is employed to capture the user's performance in using the system. Examples: Frequency measures, time measures, error measures, physiological measures.
- Observation method: Natural behavior is captured by direct observation or by the use of audio–visual instruments.
- Questionnaire: A set of written questions to be answered by the user before or after using the system, concerning past behavior, expectations, attitudes etc. with respect to the system.
- Interview: Similar to the questionnaire, but using directly spoken questions. The interview can be performed in a more or less standardized way, resulting in potentially different information which can be obtained from the user.
- Self–descriptive: A method of obtaining verbal protocols which reflect the user's thoughts or opinions, retrospective or introspective of their own behavior when using the system. The user's comments can be collected on–line while performing the task, off–line just after completing the task, or while playing back a video or audio registration of their behavior. A similar method, directed not at the usability of a system but at the user's opinion on certain stimuli, is the so–called "Intermitting Loud Thinking" method proposed by Schulte-Fortkamp (1994). It has been successfully applied to obtain information on the perceptual characteristics of car sounds, but could also be applied to evaluate the quality event when using a system, in order to get information about the relevant cues.

The choice between one or more of the usability test methods will also depend, apart from on the system or service itself, on the developmental phase in which the evaluation takes place. The earlier usability concepts are applied and evaluated during the development cycle of a product, the higher is the probability of realizing a highly usable product. In the development and improvement of voice transmission and communication systems, the usability can generally be evaluated in each phase, most commonly by the use of rating scales. However, if these experiments have the character of standardized listening–only tests, it is doubtful whether usability is the quality entity to be addressed, or more probably voice transmission quality or communication efficiency.

7. ASSESSMENT OF COST–RELATED FACTORS

It is obvious that the price of telecommunication services is an influencing factor on the quality. The price generally consists of three components: the price for the terminal equipment (e.g. the telephone set, hands–free terminal, etc.), the general access fee (subscription account) and the fee to be paid for a specific connection. All three components have an influence on what quality the user expects and how readily he/she is willing to accept a certain level of quality or degradation. Due to the temporal aspect, a user will be more influenced by the price per connection, because this is directly related to a specific connection, especially when charging information is available before, during or after the call (and not at a distant point in time when the bill is sent).

There are various ways of charging for telephone connections. In the past a fixed price–per–unit model was usually used in many European countries, and the length of a unit depended on the time of the call. Another important factor was the distance of the call (local, regional, national, international). Due to the liberalization of the telephone market, this price model has been abandoned in favor of many different models which are currently offered by service providers. Especially in mobile networks, the subscriber can sometimes choose between different charging models which are a trade–off between the account fee and the price per connection. For the distribution of charges between the caller and called party, different models exist as well. In the PSTN, it is always the calling party who is charged for establishing a connection (except special operator calls). In mobile systems such as GSM, there is mainly a split charging, where the calling party is charged for part of the connection to the GSM subscriber's home location, while the called party is charged for the roaming part of the connection. For future scenarios, such as UPT (universal personal telecommunication) or satellite bound mobile communication, other charging methods may be used, e.g. where both parties may decide on whether to accept a connection depending on the price. Charging information during or before establishing a call may also be available to the user of ISDN (integrated services digital network). The user may then decide on the general set–up and the length of the call based on the given price information.

The large variety of charging methods has an influence on how this aspect can be catered for in the quality assessment of telephone connections. Generally speaking, in a laboratory situation it is difficult to get information on how willing a user is to pay a price for a specific connection. On the other hand, laboratory experiments are mainly the only way of establishing connections with a defined transmission characteristic of a telephone line. Therefore, experiments have been performed where the users were asked in a conversation experiment about the price they would be willing to pay for each connection. Two different types of scales were used: In a first approach, the users could express the price in currency per unit (DM/unit). A standard price of 0.12 DM/unit was indicated, corresponding to the price usual at that time for the former monopolist, Deutsche Telekom. After this first approach, charging models developed in Germany, and especially experience with mobile telephony grew where other charging models came into use. In a second approach, a more abstract scale was used, asking the subjects to rate the choice in terms of "points per unit". A standard of 100 points/unit for a "normal telephone connection" was indicated. Results of these experiments and a comparison with quality judgments are given in Section 5 of Chapter 6.

As stated above, results obtained in laboratory tests do not reflect a completely natural situation, and therefore judgments on an acceptable price may be biased. Another influencing factor in laboratory situations is that test subjects tend to give lower values for the price, because they have the impression that this may influence the future development of price (especially when they know that the investigation is being done on behalf of a service provider). Therefore, it is desirable to perform user surveys and market behavior observation in order to obtain more valuable information on prices. Unfortunately, in that case price cannot be directly compared to quality.

8. SCALING

The term scaling describes the establishment of a relation between numbers and objects using specified rules. In psychophysics, these objects are perceptual in nature. The establishment of such a relation should be done in such a way that the numerical value obtained by the scaling procedure afterwards shows the desired properties (Borg and Staufenbiel, 1993, p. 214). Scaling, therefore, cannot be done in a standardized way for different types of applications. In contrast, the scaling procedure should match the aim and the object of the assessment. The very detailed discussion in this section explains why new scaling methods are worth being used in telephony. On the basis of this discussion a new scaling method for rating impairment is proposed and experimentally validated in Section 7 of Chapter 6.

In order to explain differences in scaling procedures, it is helpful to classify the different types of scales according to their properties. According to Stevens (1946), four levels of scales can be defined:

1. Nominal scale: The scale points represent identities, but no relation between the entities can be established. Example: The difficulty scale used in conversation experiments (cf. Figure 4.7).
2. Ordinal scale: In addition to the identity, a rank order between the scale values can be defined. Example: The 5–point ACR scale or so–called

'MOS scale' shows ordinal characteristics. Interval scales can sometimes be derived from ordinal scale judgments, in the case of assumptions concerning the distribution of responses and the correlation between responses being satisfied, see Guilford (1954); Sixtl (1967); Thurstone (1927).

3. Interval scale: Besides the ordinal characteristics, intervals can also be defined which have additive property. Example: Continuous scales with pre–annotated numbers or attributes can sometimes be shown to have interval level ("Method of Equal–appearing Intervals", see Guilford (1954)).

4. Ratio scale: This scale shows all the characteristics of an interval scale and has a defined zero point. Ratio scales have the property of double intensity corresponding to a double scale value. Example: Scales established by magnitude estimation can have ratio or close–to–ratio level (Borg, 1994; Borg and Staufenbiel, 1993).

The human being is good at measuring the perceptual intensity or magnitude of a stimulus. This ability refers not only to making comparative judgments, but also to some kinds of absolute judgments in the sense of inter–subjective valid responses. To do this, the human being refers to what he/she has learned and his/her experience with objects. The common knowledge about things and perception allows us to reach inter–subjective agreement, though individual perception may be different (Borg, 1992). This ability is worth noting when quality judgments have to be made which should reflect a 'general telephone user's point of view', or which lead to decisions for telephone network planning affecting a large number of users. Assessment methods should therefore reflect an 'absolute' level of perception if they are expected to be used in anything other than a purely comparative way.

The "fixed star" (Borg, 1992) which could serve as a reference in a telephone quality assessment experiment, however, is not clearly defined. If subjects are expected to make 'absolute' instead of comparative judgments, several references may have an influence on the perceived quality. Some of these so–called context effects will be discussed later in Section 1 of Chapter 6. One appropriate "fixed star" would be the 'average' telephone connection, as the user knows it from home or from his work place. In the past, this reference connection could be expected to represent a conventional, analog wirebound subscriber line. Differences present would have been the result of phone habits, such as for people that were used to making a lot of long–distance calls or international calls. Now the situation has changed, as many subscribers may now use a cordless phone or a mobile cellular phone. The implicit reference is different in those cases.

8.1 RATIO SCALING

Coming back to the scale levels defined above, it would be consequent to demand the highest possible level for telephone assessment purposes in order to obtain highly informative test results. The ratio scale would be the most appropriate scale for such a task. Ratio scaling methods have been developed by Stevens (1957), both as estimation methods and production methods. The method called 'magnitude estimation' (ME) is most widely used because of its simplicity. Test subjects are given the task to assign numbers to test objects so that the numbers reflect the relation between perceptual intensities, without any further restrictions on the range of numbers. The resulting scale is expected to have ratio characteristics. Although it has been shown that the ME technique is not ideal for producing real ratio estimates, it is still a common choice when stimulus–to–perceived–magnitude relations are assessed (Borg, 1994). It has been successfully used in a number of different perceptual assessment tasks. Zwislocki and Goodman (1980) found considerable agreement in the absolute numbers assigned to stimuli by different test teams, indicating that subjects tend to have an absolute coupling between numerals and perceptual magnitude. However, in general no absolute level information can be obtained by ME techniques, but only information about the relationships between stimuli (Borg, 1998). This is also a major drawback for applications in telephony.

8.2 ABSOLUTE CATEGORY SCALING

In order to anchor subjective ratings, scales with verbally defined labels are widely used to assess magnitudes with reference to anchored values. The most frequently used scale type in telephony is the category scale (ITU–T Rec. P.800, 1996). It is used in 'category estimation' as an interval scale (allowing mean and standard deviations to be calculated), and in 'category judgment' as an ordinal scale (allowing the number of scores within each category, median and percentile values to be calculated). For LOTs, ITU–T recommends category scales with pre–annotated numbers; for CTs and surveys ordinal scales with attributes only are recommended, which are post–annotated. However, also in CTs mean values (Y_C) are calculated, assuming that each category occupies the same interval on a perceptual continuum. Gleiss (1987a) states that category scales are closer to interval level when they are pre–annotated with numbers.

The category estimation or category judgment method producing a one–dimensional MOS value is very attractive, as it can be repeated at different locations and requires neither special test teams nor specific equipment nor knowledge of experimental techniques. It must be underlined that category rating tests have been used for more than 30 years now in the assessment of telephone circuits of various kinds with different types of degradation. Thus, it is the most established technique for measuring the quality of voice circuits, in the listening–only as well as in the conversation mode.

The ACR method, however, has a number of drawbacks which should be kept in mind, especially when test results are interpreted to justify taking one or another decision. Several of these drawbacks have been observed on many different occasions and at many locations, so that it can be assumed that the problems are of a general nature. A categorization of the effects is given in Bennett and Lincoln (1985).

1. Judges differ in their use of the category scale. This may be obvious for judgments obtained in different countries. Goodmann and Nash (1982) e.g. report on a comparative study of speech transmitted over ADPCM circuits and different reference conditions (additive and multiplicative noise) rated in seven different countries on the standardized 5–point ACR scale. They conclude that a reasonably accurate estimate of a test result at one location can be obtained by adjusting the result from another location with an additive constant. This adjustment approach suggests that a bias is present in the judgment of a test group in one country. The method for adjusting subjective ratings by transformation will be addressed later in Section 2 of Chapter 6. At AT&T, a number of listening–only tests was carried out on different low–bitrate codecs, and the results were analyzed with regard to the MOS distribution (Bennett and Lincoln, 1985). The MOS distribution, which is generally low and flat, becomes much sharper when the judges' responses are normalized to have a mean of zero and a unit variance. The differences in distribution of the MOS show that judges differ in their personal use of the category scale. Circuit conditions which in the opinion of one test subject fall into the "good" category can be associated with "poor" or even "bad" by another subject. Both observations justify the assumption that test subjects are biased measurement instruments. It is therefore not appropriate to regard the observations only as additional sources of error, which can be eliminated by using a larger number of subjects. At least a correction for the bias has to be assumed. In telephony, this is generally done by means of a transformation using some MNRU reference judgments as anchors, but the transformation is only applied to the mean rating value over all the test subjects. It will be shown later that this transformation may be very ineffective for some data and does not necessarily lead to more valid (in the sense of generalization) test results.
   The difference observed in interpreting the attributes of the scale between test subjects has a consequence for defining limits on this scale, e.g. limits for acceptable connections. In transmission planning, this limiting case is in the area of MOS = 3.35, i.e. at the border between the "good" and "fair" categories (ITU–T Rec. G.113, 1996). However, some test subjects will

only rate a normal, clear connection with the attribute "fair", or (in German) "*ordentlich*". Thus, the attributes defined for the MOS scale do not allow limits for acceptable transmission quality to be set up.

2. Context effects occur showing the interdependence of categorical judgments. This is a very strong drawback not only of the category scales but also for other kinds of absolute rating procedures. Context effects will be discussed in more detail in Section 1 of Chapter 6.

3. The perceived intervals between the attributes of the ACR scale are not equidistant. Jones and McManus (1986) investigated the position of the quantitative terms on a perceptual continuum ranging from "best imaginable" to "worst imaginable". US and Italian observers participated in the test. It turned out that the intervals were regular and the meaning varied only slightly between different geographical regions or cultures. The position of the attributes of the MOS scale as well as those of the DCR scale are reproduced in Figure 4.9. It is obvious that the intervals on the MOS scale are far from equidistant, especially for the lower categories "poor" and "bad". The same holds true for the Italian labels of the 5–point ACR scale. The attributes of the DCR scale, in contrast, are much more equidistantly spaced on the continuum. The results underline that the MOS scale does not possess interval properties. Unfortunately, to our knowledge no comparison has been made between MOS scales pre–annotated with numbers and those given without numbers for the subject's rating. Such a comparison would probably show that the pre–annotation helps to linearize the scale by putting the attributes into more equally spaced positions on the perceptual continuum. It should be noted that the non–linearity of the scale leads to a systematic error which cannot be corrected by increasing the number of test subjects, but only by using a transformation or by eliminating the error sources.

4. The sensitivity of the MOS scale is relatively low. The lack of sensitivity is mainly due to the fact that subjects are forced to use only 5 categories, and no judgments between the categories are allowed. However, many subjects often feel the necessity to make decisions between categories. The scale design leads to a loss of information which otherwise could be obtained from the subjects. It has been shown that especially untrained test subjects do not differentiate between more than 5–6 categories in some judgment tasks (McKelvie, 1978). However, if only a few categories vaguely labeled are presented, is does not become clear to the subject that a very exact rating is expected.

5. It has been observed that naïve subjects sometimes avoid the use of the extreme categories of the scale, because they expect even better or worse

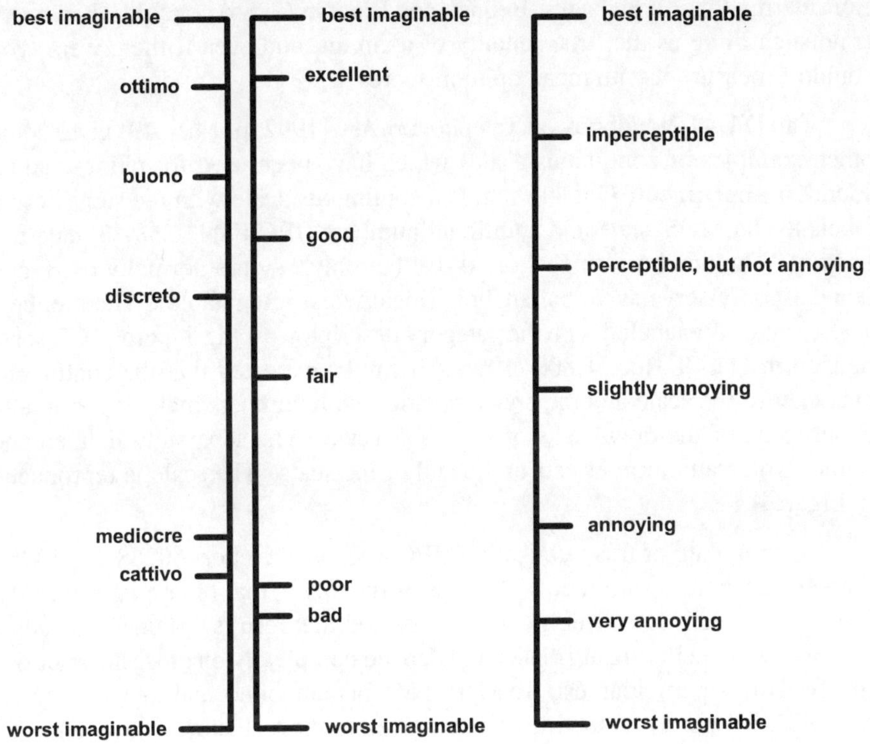

Figure 4.9. Position of attributes according to Jones and McManus (1986). Left: ACR quality scale (English and Italian version); right: DCR scale.

stimuli to follow which then could not be catered for. On the other hand, if a stimulus is rated "excellent" or "bad" and the following stimulus is perceived even better or worse, the test subject is forced to put it into the same category. The latter effect is sometimes called 'saturation effect'. Both effects can be reduced if separate scale end points are defined. Such a scale was used by Bodden and Jekosch (1996) for measuring integral speech quality and is reproduced in Figure 4.10. The Swedish Telecommunications Association used a scale similar to the one defined in (IEC Publ. 268-13, date unknown) for many years.

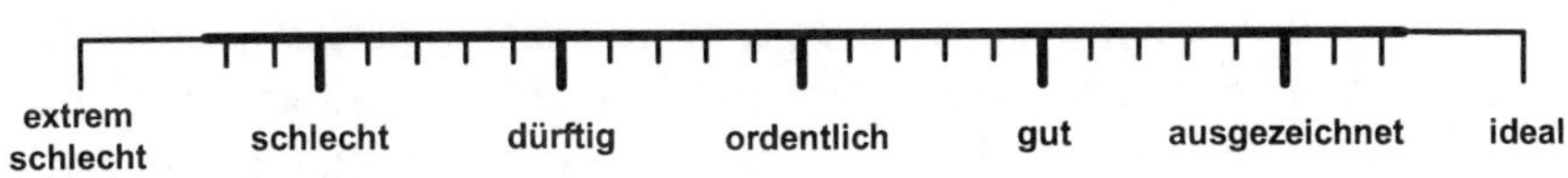

Figure 4.10. Continuous rating scale according to Bodden and Jekosch (1996), German version.

Considering these drawbacks, Bennett and Lincoln (1985) conclude that "there is no such thing as the 'true' quality of a circuit, and even if there were, you couldn't measure it with mean opinion scores".

The ITU–T Handbook on Telephonometry (1992, p. 148-149) gives some other examples of continuous scales which have been used for rating quality, sound or impairment. One of them is a continuous scale with the same verbal labels as the MOS scale and additional numbers (IEC Publ. 268-13, date unknown). The numbers range from 0–10, but only 1–9 are normally used, and 0 and 10 only serve as verbal anchors (ideal/worst imaginable). The numbers 1,3,5,7 and 9 are labeled with the category descriptors of the 5–point ACR scale defined in ITU–T Rec. P.800 (1996). It can be expected that the continuous character of the scale and the pre–annotation with numbers make it possible to avoid some of the drawbacks of the MOS scale. The separately defined end points avoid saturation effects at the end of the scale. The scale is reproduced in Figure 4.11.

In an update of this publication (IEC Technical Report 60268-13, 1998), the verbal labels are omitted. The reason for this is that "this might bias the use of the scale". However, it can be expected that results obtained on such a scale without anchoring attributes will depend completely on the test conditions involved in the particular test. To a larger extent than for normal category scales, its use also depends on the instructions given to the test subjects as well as on the first condition presented in the test. A discussion on some of the context effects observed for category scales follows in Section 1 of Chapter 6.

A correction for the nonlinearity of the 5–point ACR scale can be reached by re–assigning numbers to the categories which correspond to the positions of the labels on a perceptual continuum. As the positions may vary from subject to subject, the positions should be determined for each subject directly after the test. This approach was performed for several listening–only and conversation tests described in the following. For some tests, it turned out that the transformed ratings made slightly better use of the scale. However, for many conditions the difference was only a minimal one. The reason is that information which has not been collected from the assessing subjects during the test cannot be reconstituted by a later transformation of the results.

8.3 CATEGORY–RATIO SCALING

In order to overcome the disadvantages of category scales (low sensitivity, only interval level) and ratio scales (no 'absolute' level information, in the sense of inter–subjectively valid levels) a category scale with ratio properties was developed by Borg (1982). It combines the advantages of the ratio scale for relative level assessment with the verbal anchors and direct level estimates

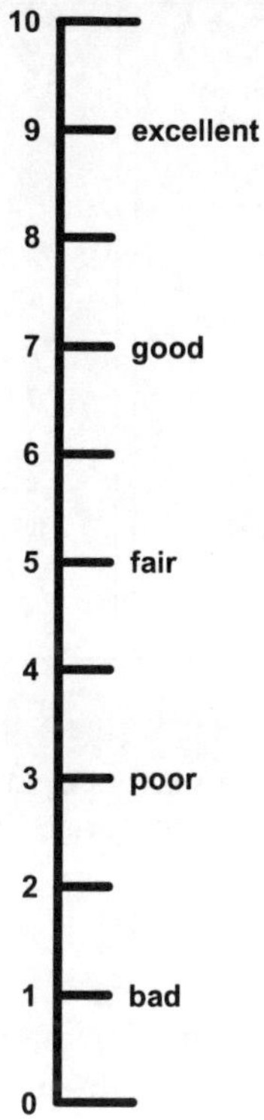

Figure 4.11. Continuous rating scale according to (IEC Publ. 268-13, date unknown).

provided by category scales. In its simplest form, the CR–10 scale (category–ratio scale), it has been successfully used in a number of different scaling tasks, namely loudness judgment, perceived exertion, pain judgment, medical applications, etc. The positions of the verbal anchors of a category scale have been changed in order to reproduce the growth function which could otherwise be obtained by magnitude estimation. The CR–10 scale is limited in numbers from 0, 0.5, 1, . . . , 10, with the possibility of using decimals or fractions as well as ratings above 10. In order to increase sensitivity at low intensities, other similar scales have been developed, e.g. CR–12, CR–60 and CR–100 scales. For practical purposes, however, the CR–10 scale generally shows enough sensitivity and is easier to use for the test subjects. The scale is copyrighted by Borg in order to avoid misuse. A reproduction is given in Figure 4.12.

Borg (1998) gives a detailed description of how to administer scaling on the CR–10 scale. Two important points have to be mentioned here. The scale value 10 (extremely strong) is regarded as the main anchor, because Borg assumes that perceptual intensities are approximately equal for different people at each individual's subjective maximal exertion. This scale value has been

0	nothing at all
0.5	extremely weak (just noticeable)
1	very weak
2	weak
3	moderate
4	
5	strong
6	
7	very strong
8	
9	
10	extremely strong (almost max)
•	maximal

0	keine
0.5	extrem schwach (gerade wahrnehmbar)
1	sehr schwach
2	schwach
3	mäßig
4	
5	stark
6	
7	sehr stark
8	
9	
10	extrem stark
•	maximal

Figure 4.12. CR–10 category ratio scale according to Borg (1982), with slight modifications of the German wording. Left: English version; right: German version. Note: The scales are copyrighted.

assigned to "the strongest perception (P) you have ever experienced". It seems difficult to translate this definition into the impairment rating in telephony. For practical applications, 10 could be assigned to "the worst telephone communication quality you have ever experienced". The results reported in Section 7 of Chapter 6, however, were obtained without this explicit anchor. Another important instruction hint is that subjects should start with the verbal expression and then choose the according number. This can easily be reached by proper subject instructions.

It should be noted that to the author's knowledge category–ratio scales have not yet been used in telephony circuit evaluation. However, they could help to improve the results of subjective experiments in the sense that the perceptual growth function is taken into account, while maintaining the 'absolute' character of the judgments. It can be applied as easily as conventional 5–point ACR scales. In our experience, subjects do not report any problems using the scale to rate impairment, which is the attribute the scale has been used for here so far. In contrast, it makes the subjects feel that a precise and valid rating is required for each rating object (speech sample or conversation), a fact which is sometimes difficult to reach with the MOS scale. The ratio property of the CR–10 scale makes it particularly interesting in rating impairment on a psychological scale, as defined by e.g. the E–model. This is a new method which is proposed and experimentally validated in Section 7 of Chapter 6.

9. DEVELOPMENT OF CONVERSATION TEST SCENARIOS

As discussed above, conversation tests are a much more appropriate way of assessing communication efficiency, as the use of the telephone is more natural compared to listening–only tests. An even more natural situation is covered in user surveys (cf. Section 5), however under the limitation of less or no control over the circuit conditions. It is therefore necessary to design conversation tests in such a way that they put the judging test subject into a highly natural situation. The key to achieve this is the design of an appropriate test scenario. Such a scenario should require only a little time for each judgment, in order to reduce the major disadvantage of conversational testing. For some purposes, conversation tests become applicable only under the condition that a certain number of systems can be covered within one test. In the following, the development of new conversation test scenarios is described with the intention of considerably reducing the conversation time while maintaining a natural and balanced dialog. The scenarios were designed and experimentally tested. They are now regarded as an economical and thus interesting alternative to standard conversation tests within the ITU–T.

9.1 REQUIREMENTS FOR CONVERSATION TEST SCENARIOS

Several types of conversation scenarios have been used in the past, e.g. the annotation of maps, the annotation of random shapes, the "stepping stones puzzles", etc. (Richards, 1973). The scenario that has been used most often to date has been the picture card set. Test subjects have to arrange six postcard reproductions of various paintings in their own order of preference, and then discuss this order with their telephone partner and come to an agreement. Common to all of these scenarios is that the conversation task is not structured, so different conversation phases (e.g. longer monologues, fast turn–taking, double–talk) have to be balanced out by using a longer conversation time. Another problem of the picture card test is that opinions have to be discussed, which may influence the subjects' opinion on the telephone transmission quality as well.

A trial to force a certain structure upon the dialog used abstract pictures (mostly from the Russian painter Wassily Kandinsky, and therefore sometimes called the "Kandinsky test"), which were annotated with a certain amount of numbers, see Kettler et al. (1994); Heinrichs (1993). Test subjects have two different annotations of the same picture. They have to compare the position of the numbers in the picture and have to find out which numbers are located at the same position. By using certain location principles, the test designer can influence the course of the dialog to a certain extent. E.g. several numbers located very close to each other may lead to a fluent discussion with many turn–takings, while only a single number at a remote location may provoke a longer monologue. The test scenario has been included in ITU–T Draft Rec. P.831 (1998) as an example.

The requirements for conversation test scenarios are multiple (Wiegelmann, 1997). The most important points can be categorized into naturalness (aspects of dependence subject $\leftrightarrow$ environment), balance (aspects regarding the course of the dialog), comparability (between scenarios for different conversations) and clarity (regarding the test instructions):

1. Naturalness
   - Natural communication task for the conversation (everyday situation, meaningful use of the telephone, no role play).
   - Natural beginning and ending of each conversation.
   - Scenario should not distract the subject's attention too strongly.
2. Balance
   - No stable roles sender/receiver.
   - Short monologues on both sides.
   - Parts with various turn–takings for both sides.
   - Provoked double–talk.
   - Balanced duration for each side.
   - Conversation as short as possible.
3. Comparability
   - Comparable instructions for all conversations.
   - Standardized dialog structure.
   - Constant dialog duration.
4. Clarity
   - Easy–to–handle instructions.

It should be noted that several of these requirements exclude each other, so only a compromise can be reached with real scenarios. In addition to the factors mentioned above, the degree of difficulty of a task may have an influence on the quality rating. Richards (1973) reports comparisons between unguided freely chosen conversation topics and a set task of comparing picture shapes. In a free conversation, a connection of identical transmission characteristics is generally rated worse. In addition, the active speech level increased in correlation with the task difficulty.

Based on the requirements given above, scenarios for a new conversation task of short duration have been developed (Wiegelmann, 1997). The topics for the conversations are enquiries and booking situations, where the telephone serves as an information exchange, a very natural purpose. The dialog structure is implied by the test instructions, which includes parts with longer monologues, others where speakers speak in turns, and some parts which should evoke double–talk. A formalization of the dialogs is given in Figure 4.13.

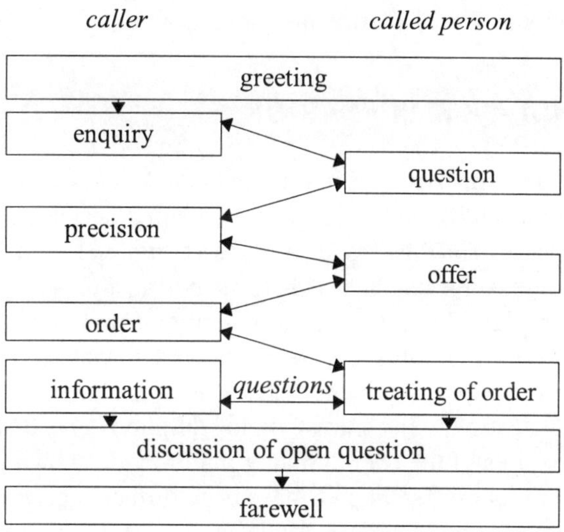

Figure 4.13. Formal structure of a dialog imposed by short conversation test scenarios.

Due to the relatively fixed structure, the dialogs may be much shorter while maintaining their balance. Various situations (railway information, travel agency, pizza service, theater reservation, doctor's appointment, flight schedule, car rental, etc.) ensure that the conversation remains interesting for the test subjects, and that different vocabulary is used. Test instructions are formalized to a high degree, using pictograms, tables, etc. As the subjects do not know which information the other requires, they cannot shorten the conversation after some dialogs (as would happen if the same scenario were repeated several times). The explanations of the dialog structure to test subjects is given in Appendix E.2, examples of such scenarios in Appendix E.1.

9.2 EXPERIENCES WITH TWO DIFFERENT TYPES OF SCENARIO

The “Kandinsky” and the enquiries and booking scenarios (in the following entitled ‘short conversation test’ scenarios, SCT scenarios) described above have been used in parallel in three conversation tests. 16 subjects were involved in each test (in all 48 subjects), and different types of impairment were investigated: room noise at the send and receive sides, codecs and echoes. The test set–up was the same for both the “Kandinsky test” and the SCT (handsets, telephone line simulator as described in Section 3 of Chapter 3, etc.), and the rating scales used were identical too. After each test subjects were given the opportunity to comment on the test. In the following, the test subjects’ comments and the observations of the experimenter are reported.

9.2.1 “KANDINSKY” SCENARIOS

Test subjects had no problems in solving the task. However, the task did not seem realistic to most of the subjects. Some of them described it as a crossword puzzle over telephone, and this was regarded as a completely unnatural situation. According to several subjects, the task became boring after some time, which is also reflected in a slight decrease of dialog duration towards the end of the test. The mean dialog duration was 332 s, with a standard deviation of 88 s. Between the eight different pictures used in the test, the mean dialog duration varied only slightly from 299 s to 357 s. In contrast, dialog duration depended mostly on the subject pair. Boxplots of the overall dialog duration are shown in Figure 4.14. Eight dialogs from four different pairs have been further analyzed regarding the active talking time for both subjects, and the percentage of double–talk time. There was no significant difference between the active times of caller (140 s) and called person (143 s). The percentage double–talk time ranged from 3.0 % to 17.6 % (mean 9.1 %).

The dialog structure depended mostly on the subject pair, and their knowledge of each other, the subject’s age, etc., but not on the picture itself. Though the location of the numbers should have provoked several dialog phases, this was only possible to a very limited extent. Several subject pairs developed a scheme for the number comparison which helped them to shorten the conversation for the last scenarios. Nevertheless, the overall balance of the dialogs seems to be quite good. To have more information on this topic, a detailed analysis of the recorded conversations is necessary, work which has not yet been accomplished. The subjects’ opinion on whether they had been distracted from the quality assessment task was ambiguous. However, the task did not seem to involve the subjects’ concentration too much.

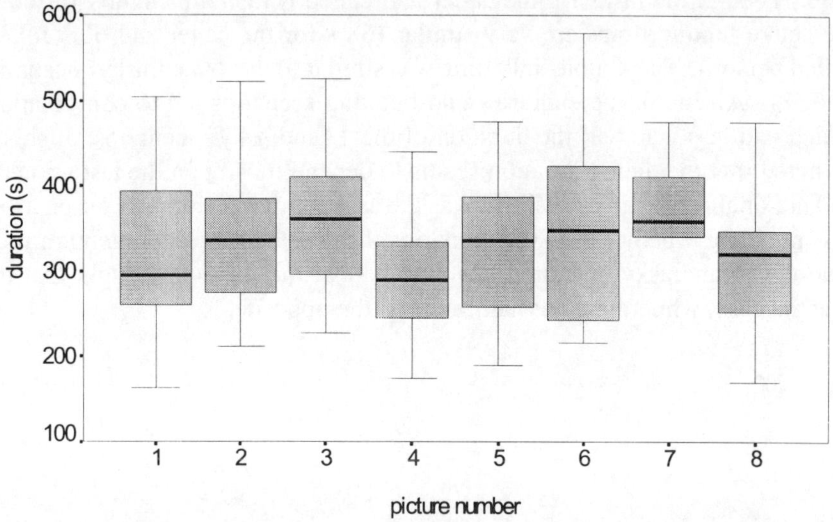

Figure 4.14. Boxplot of dialog duration for different Kandinsky pictures. Thick line marks median, grey box the interquartile range, thin lines regular minima and maxima.

9.2.2 SHORT CONVERSATION TEST SCENARIOS

Without exception all the subjects reported that this kind of scenario seemed more realistic and natural to them. As the situation was not regarded as a game, the dialogs were carried out with more consciousness, and with more 'language elements' such as vocabulary, sentence structures, *ad hoc* remarks, etc. To achieve a natural dialog behavior, subjects had to get used to the symbols and to the dialog structure implied. Therefore, at least one trial scenario should be introduced at the beginning, as was done for the tests reported below. None of the subjects found the test task very difficult, just as in a normal telephone situation. The theoretical dialog structure which was implied by the test instructions seems to work very well. Although no detailed data on the distribution of different dialog phases is yet available, the experimenter reported longer monologues for the first two parts, and many turn–takings (short questions and answers) for the following. Towards the end of the dialog double–talk phases increased. More data should be collected by making a detailed analysis of several dialogs.

The mean dialog duration was 147 s, with a standard deviation of 54 s. The range of the different scenario means was 87 to 195 s, see Figure 4.15. This range is quite large, and the longest and shortest scenario were later modified to get a more balanced dialog duration. 16 dialog conversations from 4 different pairs have been further analyzed regarding active talking time and double–talk time. Though the tasks for the caller and called person are slightly different, the active talking times are very similar (68 s for the caller and 60 s for the called person). The double–talk time was similar to the "Kandinsky" scenarios (8.6 %). All in all, the enquiries and booking scenarios led to conversations which had less than half the duration of the "Kandinsky" scenarios. It should be noted that the dialog duration seems to depend mostly on the test scenario, and not on the test subject pair (as was the case for the "Kandinsky" scenarios). It is not clear whether the SCT scenarios distract the subjects' attention more than the "Kandinsky" pictures. Around half of the subjects thought that this was the case, while the other half reported the opposite.

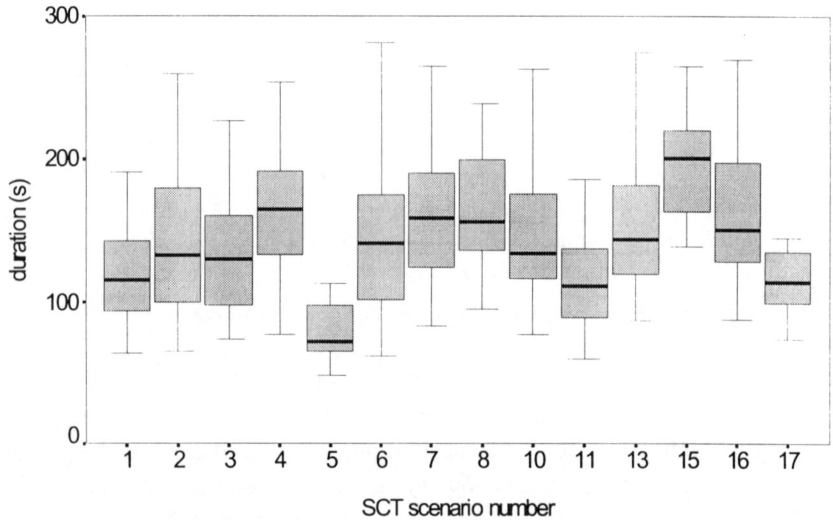

Figure 4.15. Boxplot of dialog duration for different SCT scenarios. Thick line marks median, grey box the interquartile range, thin lines regular minima and maxima.

9.2.3 COMPARISON OF TEST RESULTS

To investigate whether the SCT scenarios and the "Kandinsky" scenarios lead to comparable test results (in terms of speech transmission quality assessment), three tests regarding different types of impairment have been carried out. Each group of 16 subjects carried out both one SCT and one "Kandinsky" test session (room noise and codecs) or one SCT and two "Kandinsky" test sessions (talker echo). The results are given in ITU–T Contribution COM 12–35 (1997). For all types of degradation the MOS scores obtained agree fairly well for both scenarios. It is supposed that the differences depend mostly on the choice of test conditions within the test. Unfortunately no transformation of the MOS results could be established, because the length of the "Kandinsky" scenarios did not permit enough reference conditions (MNRU) to be included within the condition set.

9.2.4 CONCLUSIONS

Overall it can be concluded that the SCT scenarios lead to a natural and balanced dialog. In comparison to the "Kandinsky" scenarios, conversation duration can be reduced to less than a half. This very important aspect already justifies the effort put into the scenario development, because the time requirements formed a thorough limitation to using conversation tests in the past. Together with the development of a real–time simulation model (cf. Section 3 of Chapter 3), a powerful test environment is formed for carrying out valid (in the sense of the assessment aim) assessment of speech communication related quality aspects. The SCT scenarios led to similar quality judgment results compared to ordinary conversation experiments, as the tests on degradations such as room noise, codecs and echo show. The shorter duration allows testing of up to 14–15 circuit conditions within one test session of about 50–60 minutes length to be carried out. However, care should be exercised when test duration is expected to have a strong influence on the perceived quality, e.g. for time–variant degradations. In this case, it may be more appropriate to include different conversation durations in the test, e.g. by using different types of scenarios.

The SCT scenarios were developed and tested in the frame of this work at IKA, in a German version. Later, translations were made for the English and French languages. A simple translation, however, is not sufficient, because proper names and cultural habits have to be taken into account. The French version was kindly elaborated by France Telecom/CNET. This laboratory used the SCT scenarios for the auditory evaluation of echo cancellers (ITU–T Contribution COM 12–65, 1998). Mean conversation duration varied in the range of 92 s to 163 s. Compared to the scenarios used at CNET so far, a reduction in duration of the whole experimental procedure by a factor of two could be achieved. An analysis of variance, taking into account the scenario factor, confirmed that there is no significant difference due to this concomitant factor. In comparison to other scenarios, the type of scenario is insignificant for quality, annoyance, fidelity or loudness judgments.

9.3 SCENARIOS FOR SPECIAL APPLICATIONS

9.3.1 SCENARIOS FOR MOBILE COMMUNICATION

The SCT scenarios described above reflect a natural situation for wire-bound telephone connections, e.g. from office environments. For mobile environments, however, only the part of the caller (i.e. the person who would like to make a reservation or to obtain information) could be attributed to the user of a mobile phone. It is unrealistic to assume that the person providing the necessary information uses a mobile phone. For a connection between a mobile and a wirebound user, the SCT scenarios in their original version can only be used with fixed roles for caller and called party.

Additionally, it is desirable to simulate some kind of advantage of access for the mobile phone user. Only if this goal is achieved can the advantages which are inherent to the mobile terminal be realistically modeled within the test situation. The approach made here is to modify several scenario situations in such a way that the mobile phone user feels it necessary to be reachable at the specific time in the experiment. Examples:

- The mobile user receives a call from his travel agent informing him that the plane departure time has been pre–scheduled by several hours. This avoids him arriving too late at the airport.
- A theater performance is cancelled. It is rescheduled for a different date.

In this way it is possible to design meaningful scenarios for mobile phone situations, the mobile phone user being either the caller or called person. An additional practical modification is to not require the mobile user to take notes. The use of large sheets of notepaper would put a mobile user in uncomfortable situations. However, the mobile user can ask for information without taking notes, and provide only information he/she can remember easily without looking at the instruction sheets.

Scenarios designed in this way have been successfully used here for assessing the quality of a connection between a simulated mobile terminal and a conventional wirebound terminal (Plücker, 1998). The quality assessment results are discussed in Section 4 of Chapter 6. Test subjects originated from two different groups, namely 8 subjects who had experience of mobile phones because they used one, and 10 without experience. Though the simulated mobile phone had a cord and was located near the other conversation partner, especially experienced subjects (6 out of 8) had the impression of using a normal mobile phone. The remaining 2 subjects found the situation unrealistic, but mainly because the connection did not show mobile–typical impairments like time–variant noise, interruptions, etc. Around 50 % of the non–experienced subjects regarded the situation as being realistic, but most of the others criticized the good conversation quality or the closely located conversation partner. In conclusion, it can be summarized that the modifications described above result in realistic scenarios for mobile environments which show the same advantages as the original SCT scenarios.

9.3.2 TELEPHONY OVER INTERNET PROTOCOL

A modification of the original SCT scenarios has also been used in this work for conversation tests carried out using an internet terminal. The test set–up is one example for connections from a multimedia terminal, as defined in ITU–T Rec. H.323 (1998), however, covering only voice input and output capability. Similar terminals are used in all multimedia communication systems where the underlying transport is a packet based network, e.g. local area networks, metropolitan area networks, intra–networks and the internet. Common to many of these systems is that the user interface is provided by a computer screen, where a (windows based) program serves to establish an audio or audiovisual connection. The psychological effect inherent to this situation is not negligible, because the experience with such types of terminals will be different from the one for conventional telephone connections. Users of computer–based terminals may expect a voice quality similar to the normal sound capability of the system, namely wideband digital system quality. This point is further addressed in Section 4 of Chapter 6.

For the design of conversation tests carried out on such multimedia terminals (in the following simply called 'internet terminals') it is important to establish a test situation which is a realistic one for the use of such terminals. It is believed that currently an internet terminal will mainly be used when

1. the user is already sitting in front of a terminal because of some other reason, e.g. he/she is operating another program,
2. the telephone conversation is directly related to work carried out on the computer terminal, e.g. a conference with shared documents or internet shopping,
3. the telephone connection is part of a broader telemeeting conference,
4. the telephone connection is part of an audiovisual communication system, or when
5. it is much cheaper for the user to establish a connection over the internet.

The relatively poor transmission quality especially for internet based connections, and the relatively high call set–up costs will not make internet terminals attractive for establishing general purpose telephone connections. This assumption may no longer be valid in the future of evolving terminal and internet characteristics.

For the time being, items (1) to (3) assume that the computer terminal is already in use for a purpose other than to establish the telephone connection alone. Therefore it seems reasonable to impose a test situation where the terminal serves a second purpose apart from establishing the telephone connection. Audiovisual communication systems are outside the scope of this investigation; however, also for audiovisual systems one of the items (1) to (3) may be valid. A motivation purely dictated by the cost of a connection is very difficult to investigate in subjective tests under laboratory conditions, cf. Section 7. Following this reasoning, the SCT scenarios were transformed into online forms which can be established and accessed by a text–processing program. In this way test subjects retrieve and protocol the necessary information (timetables, price lists, reservation details, etc.) from the computer screen. Also the subjects' ratings are to be given in the computer online form. The topic of the conversations may, in contrast to the mobile situation, remain unchanged. Especially reservation and booking seem to be good examples for telephone connections which may be realistic for the use of an internet terminal.

A conversation test was carried out where subjects attended one session on computer terminals and one on ordinary terminals, then using the original SCT scenarios (Riedel, 1998). For the computer terminal situation, the set–up of the call originated from a commercially available internet telephony program (Cooltalk© from Netscape), where a button was provided for each connection (labeled "train information", "theater", etc.). No visual information on the communication partner was given. The set–up of the telephone line was the same for both test sessions, including the handset (which was connected via the computer terminal). The comments made by the test subjects and the observations of the experimenter showed that subjects regarded the situation as relatively realistic. The naturalness of the dialog situation was rated 3.6 on average on a 5–point ACR scale. None of the persons who already had used a computer terminal had problems using the program and providing the rates in the relevant forms. On the other hand, subjects who had never used a computer terminal sometimes experienced severe problems in operating the terminal, e.g. operating a computer mouse. These subjects, however, are not part of a potential user group, as they have no access to a computer terminal.

The possibility cannot be excluded that the subjects' ratings, in terms of 'integral quality', are also influenced by the ergonomic design of the special computer terminal they used. Especially for inexperienced test subjects the terminal operation constitutes an important factor concerning satisfaction with the system. The results of this test presented in Section 4 of Chapter 6, however, suggest that problems in operating the terminal are not specifically reflected in the integral quality rating. Inexperienced users rated the internet terminal in general higher than experienced ones. Usability evaluation of the test scenario could contribute to more information on the influence of terminal operation problems, for both experienced and inexperienced users.

10. CLASSIFICATION OF ASSESSMENT METHODS

To summarize, the assessment methods described in this chapter are classified in the QoS schematic. Apparently, all the methods which make use of a listening–only situation can only investigate features of one–way voice transmission quality. Amongst these are all the articulation and intelligibility tests, listening–only tests using absolute rating (ACR, magnitude estimation, multidimensional scaling and polarity profile), as well as listening–only tests using paired–comparison techniques (DCR, CCR, isopreference tests). This category also contains speaker identification tests, which were not discussed here. ACR tests using the listening–effort scale relate more to the ease of communication, as well as questions on intelligibility and the interference test.

Communication efficiency can only be investigated by performing full or simplified conversation tests. Two different scales are given in Section 3.5, namely the 5–point quality scale and the difficulty question. Talking and listening tests relate more to conversation effectiveness, as well as the special double–talk tests. Comfort and costs are mainly investigated by means of questionnaires. The difficulty in assessing cost–related factors in a laboratory situation was discussed in Section 7. All the methods discussed in Sections 5 and 6 relate to the usability of a service, like questionnaires, logginginxxlogging, user observation, interviews, etc. In this category the question concerning acceptability, sometimes given in a laboratory test situation ("Would you regard the quality acceptable for a certain service?"), also fits well. Real measures of acceptability, however, can only be obtained by counting the actual number of users.

11. SUMMARY

In order to qualify and quantify the contribution of different factors to the quality of a telecommunication service, auditory assessment methods are necessary. Depending on the method used, distinct types of information can be obtained from the test subjects or the users involved in the assessment. Thus, the focus of the judgment is strongly influenced by the method used, and so is the quality feature under investigation. As a result, there is no optimal test method which would provide 'best' assessment results for all test tasks. In contrast, an appropriate method has to be chosen by the experimenter according to the target of the investigation. This chapter gives an analytic overview over most of the quality assessment methods used in telephony. The existing methodologies are categorized and own developments are integrated in order to have powerful tools at hand for the experimental investigations described in Chapters 6 and 7, namely regarding test scenarios and scaling methods.

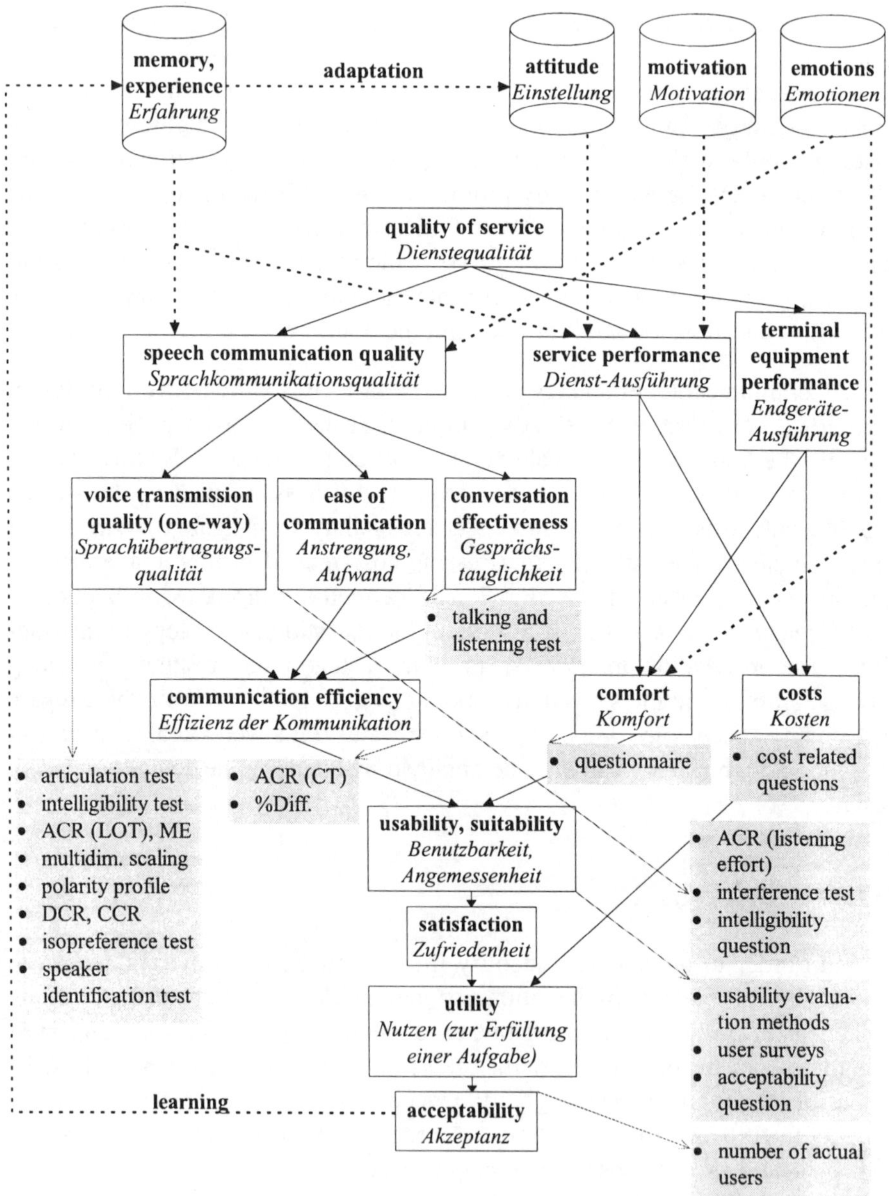

Figure 4.16. Classification of assessment methods in the QoS schematic. German terms are given in italics.

Several factors carry an influence on quality assessment. Test methods can be categorized according to the test subject group (naïve vs. experienced subjects), the scaling method, the presentation method (method of adjustment, interrogative method), the test modality (listening–only, talking and listening, conversation), the test situation (laboratory tests, surveys, field tests), etc. For a proper choice of an adequate assessment method, however, it is important to know which feature of quality, usability or utility a test method refers to. All methods making use of a listening–only situation can (inherently) address only the one–way voice transmission quality. Articulation and intelligibility tests determine the identification of certain speech elements. Because the identification is hardly influenced in ordinary telephone connections, the use of these tests is limited to connections with rather poor transmission performance. Higher quality connections can be assessed in listening–only or conversation opinion tests. Amongst the LOTs, those using absolute category rating (ACR) techniques are the most common ones. In ACR, each stimulus is rated individually on a category scale. Different scales are recommended for use in telephony, referring to integral listening–quality, listening–effort or loudness preference. More analytical information on certain dimensions of quality can be obtained from polarity profile judgments. In contrast to ACR, paired comparison techniques are in general more sensitive to little differences between stimuli. These techniques, however, do not permit to determine 'absolute' quality levels, in the sense of the relationship to an external reference for the assessing test subject (e.g. to the usual telephone connection at home). The same holds true for ratio scaling techniques such as magnitude estimation.

To investigate communication efficiency, it is necessary to carry out experiments in a natural conversation situation. CTs do not only permit to investigate transmission characteristics which affect the talking or talking–and–listening situation (e.g. echoes, overall delays, etc.); they also better reflect the 'absolute' quality horizon than listening–only tests do, and thus provide more valid test results. Inherent to the test situation several factors which may have a strong influence on LOT results are balanced out, namely variability between speakers regarding voice characteristics, speech level, speaking style, etc. The main drawback of conversation tests are the high costs in terms of the time and money needed to carry out the experiment. This is a severe limitation which sometimes renders CTs unusable. To overcome the problem, test scenarios were developed which permit the experimental time to be reduced by a factor of 2–3. The scenarios were specifically designed to be natural (reflecting the normal relation between subject and environment), balanced (regarding the course of the dialog), comparable (between different conversations) and clear in their instructions. The topics of the SCT scenarios are enquiries and booking situations. Experimental evidence is given that the SCT scenarios lead to adequate conversations of moderate duration and maintain the quality focus of natural conversations, however reducing the experimental effort. For special applications other than wirebound telephony, modifications were developed and tested. These modifications reflect the requirements for mobile phone situations and for phone calls carried out from a computer terminal.

Proper scaling methods have to be chosen if highly informative descriptions of quality events are required. Therefore, the advantages and disadvantages of usual scaling techniques were analyzed in detail. It turns out that the usual category rating technique is more adequate for determining 'absolute' levels. The common knowledge about things and perception allows to reach inter–individual agreement though individual perception may be different. This ability is better captured by verbal category labels. However, the 5–point ACR quality scale (the so–called MOS scale) commonly used in telephony has several drawbacks which partly result from the scale characteristics. A higher ratio scale level can be obtained in ratio scaling experiments. In this way, however, only information about relationships between stimuli is captured. The category–ratio scaling method according to Borg (1982) allows to combine 'absolute' level information with a high (ratio) scale level. This scaling method will be used later for the assessment of impairment as defined by one of the transmission rating models, see Section 7 of Chapter 6.

Tests under laboratory conditions suffer from the lack of naturalness. This is due to the test subjects giving their opinion in a test situation, and carrying out conversations which are — even in naturally designed CTs — not really useful to them. In order to assess usability, methods are described which focus on the function a system has for its user. These methods mainly consist in observing the user of a system or asking his/her opinion by means of questionnaires or interviews. User surveys allow usability and even utility related features to be captured. On the other hand, not very detailed information on certain characteristics of voice transmission quality or communication efficiency can be obtained (as it would be possible in LOTs and CTs). One major drawback is that no direct relationship between actual quality opinion and a certain network configuration can be established in most cases, due to the lack of control over the circuit conditions. An experimental system called SYBIL overcomes this limitation, however restricted to the business call domain. The assessment of cost–related factors is also difficult under laboratory conditions. Acceptability measures require the investigation of the number of actual users and are therefore limited to field tests and actual system investigations. This underlines that a wide choice of test methods is necessary in order to capture the different dimensions contributing to the quality of a service. As far as it is possible under laboratory conditions, the described methods are used in the following experiments (Chapters 6 and 7) to investigate the contributions of different quality dimensions.

Chapter 5

MODELS FOR PREDICTING SPEECH COMMUNICATION QUALITY AND SERVICE–RELATED MODELS

Telecommunication networks, either public or private, have to be planned in order to satisfy the users' needs. Their characteristics have to be adapted to the users' behavior in the communication situation. The quality the user experiences when using the system can ultimately be investigated only by observing user behavior and asking for users' opinions. Methodologies for that purpose were discussed in the last chapter.

In the planning process, however, it is often too expensive and time–consuming to set up simulations of potential future network configurations and to carry out auditory experiments using test subjects. In order to facilitate the planning process, data on auditory tests carried out at different locations were collected in a systematic way from the 1960s onwards. These data can already provide very useful information on the judgment of auditory events diverse in character. In view of the multitude of different types of impairment encountered in modern telephone connection scenarios, however, it is not realistic to cover user judgments by tabulating experimental data. Therefore, it was necessary to develop theoretical models in order to relate users' judgments on telephone speech quality to instrumentally measured or reference data of the telephone connection of interest. Several of these models were developed independently by telecommunication administrations, assuming different kinds of input parameters to describe the telephone transmission configuration, and different types of output data. Four different models were finally considered by the ITU–T for use in transmission planning, and an algorithmic description can be found in ITU–T Suppl. 3 to P–Series Rec. (1993). These models take into account most of the degradations occurring in traditional analog networks, e.g. loss, attenuation/frequency distortion, circuit noise, ambient (room) noise, talker and listener echo, or quantizing distortion resulting from pulse code modulation. They disregard, however, some important distortions introduced by modern digital networks, especially the effects of low–bitrate codecs. During the early 1990s, work was started in order to integrate the experience gained in using the models to set up one new model which includes the effects related to low–bitrate codecs. This model was developed by a group within ETSI (Johannesson, 1997) and is called the E–model. It has been standardized both by ETSI (ETSI Technical Report ETR 250, 1996) and ITU–T (ITU–T Rec. G.107, 1998), and is now considered to be the only model that should be used in the transmission planning of modern networks.

It is not the aim of this chapter to describe all the existing models in algorithmic detail. Instead it is interesting to discuss the different underlying principles in order to understand how relations between different types of impairment are established. The models can be broadly separated into two groups based on their input parameters: models which make use of frequency–dependent quantities and those which use single, scalar parameters. All the models include major components which were derived using empirical data. Therefore they cannot be expected to give valid information on the perception process itself. However, especially the SUBMOD model discussed in the next section is partly based on theoretical considerations on human perception. The validity of the resulting quality predictions will be discussed in more detail in Chapter 7.

The transmission planning models discussed above take into account the whole transmission chain from mouth to ear. Additional models, sometimes called instrumental or even (inappropriately) "objective" measures, try to predict quality deterioration due to certain parts of the transmission chain, namely codecs. These models do not use measurable quantities of the transmission chain as input parameters, but input and output signals of the piece of equipment under study. They can therefore not directly be used in transmission planning. However, predictions of instrumental measures can be used to optimize transmission planning models and to provide input parameters, in the case that for a newly developed piece of equipment (e.g. a new codec) no information is available on its impact on transmission quality. A brief discussion of the instrumental measures is given in Section 3. For a more detailed analysis the interested reader should refer to the recent literature (Berger, 1998; Hauenstein, 1997; Quackenbush et al., 1988).

Apart from the models which relate to speech transmission, there are some approaches for modeling the effects of problems in call set–up and call completion. The related effects were much more important in the early days of analog transmission without bandwidth compression. However, in view of the behavior of packet–based transmission systems like the internet, which are expected to carry part of the voice communication load in the near future, it is worth having a new look at such modeling approaches. Moreover, models of strong user reactions (like abandoning a call or complaints to the network operator) should be seen in this context. These models are discussed in Sections 4 and 5.

1. MOUTH–TO–EAR MODELS USING PARAMETERS IN THE FREQUENCY DOMAIN

According to the classification introduced by British Telecom (ITU–T Suppl. 3 to P–Series Rec., 1993, p. 16), there are two types of models which predict the performance of complete telephone connections mouth–to–ear. One is based on pure empirical modeling of subjects' opinions on test connections, using mainly curve fitting algorithms. Models established in this way have no further significance as their relations do not explain or model the human perception of a quality event. Two of these models are explained in further detail in Section 2.

On the other hand, there are models which are structured in such a way that reflects the perception process as a cause–and–effect relationship between input parameters (sensitivities of the speech path, acoustic environment, listeners' hearing characteristics, emitted speech characteristics, etc.) and an output parameter, e.g. the users' opinion of performance. There are three models of this second type discussed in ITU–T Suppl. 3 to P–Series Rec. (1993), namely the Information Index developed by France Telecom CNET, NTT's OPINE model, and the SUBMOD or CATNAP model developed by British Telecom. It is believed that these models would be more flexible for modeling characteristics which were not part of the material used for the establishment of the model. However, similar to the first class of models, at some point all the models make use of empirically determined functions to form quality estimates, because the quality event cannot be described by parameters or functions of the human auditory system. In turn, also the scalar parameter models implicitly take characteristics of human perception into account, namely loudness ratings and frequency–weighted noise levels. Thus all transmission planning models contain an empirically determined part. This is not the only element which limits the validity range of the model.

In the following, the SUBMOD model developed by British Telecom will be discussed as an example of models of the second type. It is mainly based on the loudness rating model described in Section 1.1 of Chapter 3. A first description can be found in Richards (1974), and the recommended version (then called CATNAP) is explained in Section 2 of ITU–T Suppl. 3 to P–Series Rec. (1993). The model has been extended to include the effects of talker echo, but these modifications have not been published yet. In the form described here, the model uses an average speaker spectrum and hearing threshold, the sensitivity of the transmission path (S_{MJ} and S_{JE}), circuit noise and room noise spectra, and sidetone path characteristics as input parameters. The model calculates opinion estimates in two stages: first, an estimation of the listening–effort score Y_{LE} is derived. In a second step, this score is converted into a conversation opinion score, Y_C.

The starting point of the model is the loudness rating model, as illustrated in Figure 3.2. Unfortunately, neither loudness nor articulation correlate well with the listener's opinion on overall quality. However, use can be made of the integration over the 'eye' depicted in Figure 3.2. Instead of loudness (Equation 3.2), a quantity termed the 'listening opinion index' is calculated, using a similar function, but amended by two multiplying factors A_{LOI} and D_{LOI} that can range from 0 to 1:

$$LOI = A_{LOI}\, D_{LOI} \int_0^\infty P(Z)\, B'_{LE} df \qquad (5.1)$$

The factor A_{LOI} depends on the loudness of the received speech, and is equal to unity at the preferred listening level (this level is chosen to correspond to $OLR = 8$ dB). The exact form of A_{LOI} is illustrated in ITU–T Suppl. 3 to P–Series Rec. (1993). D_{LOI} is a function of the level of circuit noise and therefore related to the lower boundary of the 'eye'. It becomes significantly less than unity if circuit noise is heard over the line.

The listening–opinion growth function $P(Z)$ is similar though not identical to the one used in articulation, in the sense that it ranges from 0 to 1, but reaches its upper asymptote more gradually, cf. Figure 3.3. The exact formula is

$$P(Z) = \begin{cases} 10^{\frac{Z+3.8}{10}} & \text{if } Z < -11 \text{ dB} \\ 1 - 10^{\frac{-0.3(Z+14)}{10}} & \text{otherwise} \end{cases} \qquad (5.2)$$

Because $P(Z)$, A_{LOI} and D_{LOI} range from 0 to 1, LOI reaches unity only for wideband, noise free connections at an optimum listening level. The frequency weighting function is slightly different from the one derived for loudness, see Figure 3.4. In practical applications, B'_{LE} and $P(Z)$ are determined in frequency bands, and Equation 5.1 is replaced by a summation

$$LOI = A_{LOI}\, D_{LOI} \sum_{i=1}^{N} P(Z_i)\, B'_{LE,i} \qquad (5.3)$$

The listening opinion index is related to Y_{LE} in a manner which reflects the standard of transmission the users have been exposed to in their experience. If the best circuit condition included in a listening–only test is a clear, wideband connection, Y_{LE} may reach its maximum value; however, if the best condition included in a test is the standard telephone connection encountered in real wirebound networks, the maximum of Y_{LE} will not be observed under any test condition. To cater for this effect, a parameter called LOI_{LIM} is included in this relation, which corresponds to the maximum Y_{LE} observed in an experiment. It is generally set to 0.885 when connections are being judged on against the background of the British Telecom network (ITU–T Suppl. 3 to P–Series Rec., 1993). The relationship between LOI and Y_{LE} is given by

$$ln\left(\frac{Y_{LE}}{4-Y_{LE}}\right) = 1.465\left[ln\left(\frac{LOI}{LOI_{LIM}-LOI}\right) - 0.75\right] \quad (5.4)$$

In this way an estimation of the listening–effort opinion score is derived, which still depends on the listening level by means of the speech spectrum characteristics included in the calculation of the sensation level Z.

The SUBMOD model can be used to convert listening–effort opinion scores Y_{LE} into conversation opinion scores Y_C. This relation is of general interest for the comparison of listening–only and conversation opinion scores, and experimental data will be discussed in Section 3 of Chapter 6. The underlying assumption is that the vocal level of the speaker will deviate from the one assumed as the input into the listening–only test speech data. Participants in a conversation tend to raise their voices if incoming speech is low in level or poor in quality, and lower their voices when incoming speech is loud. Additional changes in vocal level will be introduced by a loud sidetone signal. Designating V_C the vocal level in conversation and V_L the vocal level in listening–only condition, V_C at end A depends on Y_{LE} at end A, which depends on V_C at end B, which depends on Y_{LE} at end B, which depends on V_C at end A, etc. In this way a circular dependence of the vocal levels in conversation is established.

In the SUBMOD model, this circular dependence can be described by the two formulae

$$\begin{aligned} ln\left(\frac{Y'_C}{4-Y'_C}\right) &= 0.7\Bigg[ln\left(\frac{Y_{LE}}{4-Y_{LE}}\right) + 0.5 \\ &\quad - \frac{K\cdot(13-STMR)}{20}\left(\frac{4-Y_{LE}}{Y_{LE}}\right)^2\Bigg] \end{aligned} \quad (5.5)$$

and

$$V_C - V_L = 4.0 - 2.1\,Y'_C - 0.3\,K\,(13-STMR) \quad (5.6)$$

where

$$K = \begin{cases} 1 & \text{if } STMR < 13 \text{ dB} \\ 0 & \text{otherwise} \end{cases} \quad (5.7)$$

and Y'_C is an intermediate quantity, which can be transformed to Y_C. The course of the calculation is as follows: By substituting in Equation 5.4 the value of Y_{LE} calculated for end A (which would be valid for $V_C = V_L$ at end B) a first approximation of Y'_C is obtained, then from Equation 5.6 an approximation of V_C at end A. The calculations are repeated for side B using this new speech level, which gives an approximation of Y'_C and V_C at end B. The circle is continued until no more larger changes can be observed, so that both equations are satisfied for both sides, A and B. Then, the relation

$$ln\left(\frac{Y_C}{4 - Y_C}\right) = 0.8451\, ln\left(\frac{Y'_C}{4 - Y'_C}\right) - 0.2727 \qquad (5.8)$$

gives the final estimate of Y_C. The relation between Y_C and Y_{LE} will be further discussed in Section 3 of Chapter 6.

Note: In the SUBMOD model, both listening–effort and conversation opinion scores are expressed on a scale from 0 to 4. The corresponding values in the normal range 1–5 can be obtained by simply adding unity to the final results.

The SUBMOD model can also take into account degradation due to signal–correlated noise, as is introduced by PCM quantization. This is done by means of a relation transforming signal–correlated noise of a signal–to–noise ratio of Q dB to a continuous circuit noise of level G dB, found to have the same subjective effect in listening–only tests with unknown sentence material and the listening–effort scale (Richards, 1973, p. 296):

$$G = 1.07 + 0.258\, Q + 0.0602\, Q^2 \qquad (5.9)$$

As stated above, the SUBMOD model has been extended to include the effects of talker echo, but this modification has not been documented yet. A software simulation of the program was available to the author, and this has been used to calculate model predictions used for comparison in Chapter 7.

2. MOUTH–TO–EAR MODELS USING SCALAR PARAMETERS

Whereas the SUBMOD model described above started from a user perception point of view (using sensation level, hearing threshold, etc.), the scalar parameter models mainly present empirical curve fits to a larger amount of data which was available to the respective model developers. In this way they can describe the effects of transmission impairment on user perception of quality (and that sometimes with a high degree of validity), but they cannot explain why the relationship is shaped in a certain way. It cannot be proven that the model is correct or incorrect, but only limited comparisons of actual user opinion and model predictions can be established. Therefore, it may be very dangerous to extrapolate model predictions on a range (e.g. of input parameters) which is not within the set of data used for its construction. Nevertheless, the models can be especially useful to network planners, because very often scalar parameters, either recommended or measured values, are the only information available at that stage of network construction. The applicability limits of transmission planning models which make use of scalar parameters will be pointed out by comparison to auditory test data, see Chapter 7.

The first model to use scalar parameters was the Bellcore transmission rating model (BcTR). It was based on extensive tests, and was first developed for the effects of circuit noise, OLR and talker echo (Cavanaugh et al., 1976). Later, it was extended to include quantizing distortion, listener echo and sidetone. The model will not be discussed in detail, because many of its concepts have been included in the E–model, which is now the only recommended model for use in transmission planning. The algorithmic description of the BcTR model is given in ITU–T Suppl. 3 to P–Series Rec. (1993).

Both BcTR and the E–model make use of the concept of a so–called 'transmission rating scale'. The reason for introducing this scale is the observation made that subjective test results are to a large extent influenced by factors related to the subject group, the type of test, and the range of test conditions. These factors cause changes in the subjects' mean opinion in terms of MOS. The introduction of a second, "psychological" scale should separate the relation between a general 'transmission rating' and the circuit characteristics on the one side, and the relation between subjective ratings and the imaginary 'transmission rating' on the other. The first part was expected to be mainly independent of individual tests, whereas the second relation could be established for each single test. Because there is no standard method for directly measuring the 'transmission rating', this assumption cannot be proven. An approach to measure a quantity which has at least characteristics similar to the transmission rating is discussed in Section 7 of Chapter 6.

The transmission rating scale, as introduced by the BcTR model, is fixed by two circuit conditions defined by their corresponding OLR and circuit noise parameters. The corresponding E–model transmission ratings R can be calculated for the same input parameters, assuming the other parameters set to their default values defined in Table 3 of ITU–T Rec. G.107 (1998), cf. Table 5.1. It can be seen that the anchor points for the transmission rating scale are essentially the same, so the BcTR transmission rating values will be comparable to the ones from the E–model.

The E–model uses the same basic concept as the BcTR model. The (physical) transmission parameters are combined to form a transmission rating factor R, which later can be transposed into estimates of users' satisfaction, e.g. the

Table 5.1. Comparison of BcTR transmission rating and E-model R for the two anchor points defined in ITU–T Suppl. 3 to P–Series Rec. (1993), Section 1.1.

OLR (dB)	N_c (dBm0p)	BcTR transmission rating	E–model transmission rating
16	-61	80	82.6
31	-46	40	37.2

percentage of users rating a connection good or better (%GoB), or poor or worse (%PoW), or into estimates of users' opinion in terms of MOS. The E–model gives estimates for the whole transmission mouth–to–ear, and for the conversation situation (although listening–only tests were used for its derivation as well). It is important to note that separate estimations of user opinion are calculated for each side of the connection, which may be different for unsymmetrical connections. The basic E–model principles are discussed in Section 2.2, and the complete algorithm is given in Appendix G.

2.1 THE ADDITIVITY PROPERTY OF THE E–MODEL

One fundamental advantage of the E–model is that it integrates all types of impairment which have been handled by network planning models, plus it covers low–bitrate codecs as a new feature. The underlying principle for handling so many different types of impairment has already been given in the description of the NTT model OPINE: "The model's basic principle is the fact that evaluation of psychological factors (not physical factors) on a psychological scale is additive" (ITU–T Suppl. 3 to P–Series Rec., 1993, p. 36). This principle is based on observations made by J.W. Allnatt and coworkers in 1964–65 on subjective quality ratings of impaired video pictures. The aim of their study was to find a method "to predict the overall quality grading when several unrelated types of impairment exist simultaneously, given the grading for each impairment separately" (Lewis and Allnatt, 1965). Their approach is, in first instance, empirical in nature. If one assigns p_0 the mean normalized category opinion score for the residual impairment ($p \in [0;1]$), and $p_1 \dots p_n$ the mean normalized opinion scores for n unrelated impairments taken separately, then the mean normalized rating p_s for all impairments occurring simultaneously is given by

$$\frac{1}{p_s} - 1 = \sum_{r=0}^{n} \left(\frac{1}{p_r} - 1 \right) \tag{5.10}$$

The formula has been empirically found for monochrome pictures with two unrelated impairments, namely random noise and single long–delayed echo.

As a psychophysical basis, it is argued that the mean normalized opinion score p could be seen as analogous to the success of an event. Then, the summable quantity $\frac{1-p}{p} = \frac{1}{p} - 1$ could be interpreted as the odds against success. As a consequence, the formula states that the overall 'odds against' are equal to the sum of the 'odds against' for the individual impairments.

In a later publication Allnatt (1975) changes from the rating on a category scale to the rating on a continuous finite scale ranging from zero to unity. The level of sensation on this scale, t, is postulated to be related to the stimulus magnitude d by the logistic function

$$t = \frac{1}{1 + |\frac{d}{d_M}|^{-g}} \tag{5.11}$$

d_M is the midopinion value of d for which $t = 1/2$, and is four times the midopinion slope of the logistic function. This function translates the whole range of possible opinions to the range $t \in [0; 1]$. On the other hand, Fechner's basic law relates perceptual magnitude or intensity N_P to the logarithm of physical stimulus magnitude. Stevens and coworkers examined a number of different modalities (Stevens and Galanter, 1957). They found that stimulus magnitude and perceptual magnitude relate to each other by a power function, the so–called 'power law'. In its simplest form without additive constant (Stevens and Galanter, 1957), it can be expressed by

$$N_P = a \cdot d^{\,b} \tag{5.12}$$

with a being the normalizing constant for the stimulus magnitude and the exponent b characterizing the particular kind of stimulus. The equation can be rewritten as

$$d = \left(\frac{N_P}{a}\right)^{1/b} \tag{5.13}$$

By substituting d in Equation 5.11 one obtains

$$t = \frac{1}{1 + \left(\frac{N_P}{d_M^b a}\right)^{-(g/b)}} \tag{5.14}$$

The constant $d_M^b a$ can be chosen to have $N_P = 1$ when $t = 1/2$, thus

$$t = \frac{1}{1 + N_P^{-k}} \tag{5.15}$$

with $k = g/b$. This equation establishes a simple relation between finite continuous scale rating (with category rating as a special case) and ratio scaling.

Empirical evidence is given that k is usually in the range [0.7;1.5]. Supposing $k = 1$, the perceptual magnitude of a stimulus on a ratio scale N_P can be expressed as

$$N_P = \frac{t}{1-t} \tag{5.16}$$

Now, t is supposed to present the subjective rating of some kind of impairment, i.e. t has the opposite quality assignment to p, or $p = 1 - t$. In that case the left side of Equation 5.10 would correspond to a term $t/(1-t)$ on the continuous scale for rating impairment. On this scale the additivity property expressed by Equation 5.10 could be expressed by

$$\frac{t}{1-t} = \sum_{r=0}^{n} \frac{t_r}{1-t_r} \tag{5.17}$$

or, using Equation 5.16,

$$N_P = \sum_{r=0}^{n} N_r \tag{5.18}$$

Allnatt (1975) states that the law of addition tends to be more accurate when expressed in terms of t rather than in terms of p, because of the difference in the basic shape of the logistic function due to quantizing in categories.

Equation 5.18 states that the impairment ratings obtained by magnitude estimation on a ratio scale are additive if multiple unrelated impairments are rated, in the sense that the perceptual magnitude of the sum of all the impairments equals the sum of the magnitude ratings of the individual impairments. It should be kept in mind that the empirical relation 5.10 was found for impaired video pictures. The concept of additive apparent magnitude is somehow contrary to the multidimensional nature resulting from unrelated impairments. The multidimensional approach would suggest a magnitude of combined impairments to be equal to the root of the sum of squares, if both dimensions are orthogonal axes in the multidimensional perceptive space. One explanation would be that the ratings investigated here concern 'impairment' and not 'quality'. It seems reasonable to assume that impairment, in the sense it is used in (audio and video) telecommunications, is a more unidimensional attribute than e.g. overall quality. One could expect that a supplementary impairment introduced to already impaired stimuli would have a smaller impact than the same amount of impairment had on a clean stimulus. This nonlinear relationship is already covered in Equation 5.16. The concept of apparent magnitude introduced here seems to provide a better explanation of the additivity than the 'odds against' principle cited above.

The question arises as to why impairment in audio as well as video applications is mainly assessed by means of category rating experiments, i.e. using a scale where the additivity does not seem to be given. In the discussion of scaling methods, some drawbacks of the magnitude estimation technique are given which can serve as a response to this question. In that case, however, the category–ratio scaling method could serve as an appropriate means of assessing telephone speech which can be degraded by several types of impairment. The approach is discussed in more detail in Section 7 of Chapter 6.

2.2 DESCRIPTION OF THE E–MODEL

The transmission rating factor R is composed of five terms, which subsume different types of impairment in an additive way:

$$R = Ro - Is - Id - Ie + A \tag{5.19}$$

The components of this equation will be discussed in more detail in the following.

Ro presents, in principle, the transmission rating for the basic signal–to–noise ratio at the virtual 0 dBr point of the connection. It has been derived from the BcTR model, and is calculated taking into account the attenuation of the transmission path (SLR and RLR) and all the possible sources of noise (except quantizing distortion noise). The noise sources from continuous circuit noise, an equivalent of the room noise at the send and receive side, and the so–called noise floor $Nfor$ are combined by power addition to the total noise, No. $Nfor$ was defined to be the induced noise on subscriber lines, e.g. hum of the fundamental frequency and its harmonics. However, it is now expected to also represent some property of human hearing, which is not affected by noise below a certain threshold. One question remains unanswered, namely whether the (fixed) level of $Nfor = -64$ dBmp has been set too high, seeing that this level is very audible in auditory experiments and seems to lead to quality degradations.

The terms I are the so–called ‘impairment factors’ which represent additive quantities related to ‘impairment’ or degradation on the transmission rating scale. The original meaning of the term ‘impairment’ was related to an equivalent change in loss, determined by a comparison with a reference system (Richards, 1973). It is now used in the sense of degradation. The degradation of each piece of equipment can be transformed to an impairment factor value. It is expected that this value is an adequate measure of the degradation introduced by this piece of equipment, and that the impairment factors are additive on the scale of the transmission rating factor R. Further details of the methodology of impairment factors is given in ITU–T Rec. G.113 (1996).

Is represents degradations which occur simultaneously with the voice signal, especially a too loud voice signal itself (a too quiet signal is covered by Ro), non–optimum talker sidetone, and quantizing distortion. Quantizing distortion is the general term used for degradations introduced by A/D and D/A conversions, logarithmic PCM coding (A– or μ–law), ADPCM coding, and conversions between PCM and ADPCM. It is modeled in exactly the same way as in the SUBMOD model, cf. Equation 5.9. The optimum OLR is a function of the overall equivalent noise No, recognizing the fact that for high noise levels a low OLR (loud connection) is an advantage rather than a disadvantage. The function for calculating the impairment due to non–optimum talker sidetone also takes $TELR$ into account, i.e. the loudness of talker echo. It can be assumed that some masking is considered by this interdependence.

The degradations which occur delayed with respect to the voice signal are taken into account by the impairment factor Id. These are impairments due to talker echo and listener echo, and communication difficulties due to a long overall transmission time (absolute delay of the connection). Some masking is considered for talker and listener echo, mainly due to the overall noise (No), and to the sidetone for talker echo. The effects of overall delay are modeled with an emphasis on highly interactive conversations and one–way delays of $Ta \leq 800$ ms (ETSI Technical Report ETR 250, 1996).

The new feature which makes the E-model a very attractive tool for planning modern networks is the equipment impairment factor, Ie. It should take into account all the impairments caused by more complicated, new equipment. Currently, it is used mainly for predicting the effects of low–bitrate codecs on the overall quality judgment in terms of MOS, and this will be further discussed in Section 3 of Chapter 7. Future extensions have been proposed, e.g. to include the effects of bit–errors and even packet loss in packet based transmission systems. However, bit–errors and frame or packet loss will have different effects, depending on the type of codec, frame length, etc. Common to these approaches is that the factor Ie is used for modeling because it can be calculated simply from single auditory test results, by a transformation rule standardized in Annex E of ITU–T Rec. G.113 (1996). This transformation involves an indirect comparison with a reference distortion from an MNRU, which is very different in type from the degradation to be described. Therefore the approaches should be examined critically. Their validity will be very restricted.

A somewhat more elaborate approach is to characterize the effects introduced by echo cancellers in terms of an equipment impairment factor (Johannesson, 1996). Some of the degradations only occur during part of the conversation, namely in single–talk (when only one person speaks) or in double–talk situations, when both conversation partners speak at the same time. In recognition of this fact a rough time weighting is applied ($Iecw$ denotes the weighted impairment factor, and Iec the one determined according to the standardized procedure, when only the degradation under consideration is present): for impairments occurring only once or twice during a conversation $Iecw = Iec/4$, for impairments occurring only under double–talk situations $Iecw = Iec/2$, and for impairments occurring in single and double–talk situations $Iecw = Iec$. This simple weighting function has not been verified, and it is believed to not be able to cater in a precise way for the perceptive characteristics which accompany echo cancellers. E.g., for degradations occurring only once or twice during a connection, it will be important whether they occur at the beginning or at the end of the conversation. However, this is an example of how the simple equipment impairment factor principle (integrating over all characteristics) can be modified and extended for new types of equipment. It shows that the E-model is not a static program, but rather a dynamic model which might be further extended — provided that the congruence with auditory test results can be proven.

The last of the additive components resulting in the transmission rating R (Equation 5.19) is the so–called expectation factor A. The letter A is an abbreviation for the 'advantage of access' which certain systems have over conventional wirebound telephone systems. A mobile communication system is an example. A user may take full advantage of being able to communicate from different locations and to be reachable, in that this compensates for a lower voice transmission quality which may be inherent to such systems. A similar situation is a multi–hop satellite connection to hard–to–reach regions. Provisional values for A are given in Table 5.2. The expectation factor A has been provisionally chosen to compensate for half of the special degradation which is inherent to such a system, in terms of impairment factors. E.g., it compensates for half of the equipment impairment factor for a GSM-FR codec in mobile situations ($Ie = 20$; $A = 10$), or for half the impairment introduced by long overall delay in the case of multi–hop satellite connections. It is obvious that the correction achieved by A is only a very rough one. Especially, it does not take into account the special situation of the user which it pretends to model. A further discussion of the expectation factor will follow in Section 4 of Chapter 6.

E–model transmission ratings R can be transformed into user reactions, i.e. MOS, %GoB or %PoW, using the normal distribution function, as explained in Appendix G. The expressions for %GoB and %PoW have been chosen according to data given in the description of the BcTR model. They present a mean over three databases, namely tests using the SYBIL method (see Section 5 of Chapter 4), CCITT conversation tests and interviews with North American customers following a long toll connection. A transformation to overall quality opinion in terms of MOS mainly follows the relations between %GoB, %PoW and MOS, as given in Annex A of ITU–T Rec. P.11 (1993). However, recog-

Table 5.2. Provisional values for the expectation factor A.

communication system example	maximum value of A
conventional, wirebound	0
mobility by cellular system in buildings (e.g. DECT)	5
mobility in a geographical area by vehicle (e.g. GSM)	10
multi–hop satellite connections to hard–to–reach regions	20

nizing that test subjects rarely use the upper end of the 5–point scale for rating telephone quality, model predictions reach saturation at 4.5, thus MOS $\in$ [1 ; 4.5] instead of MOS $\in$ [1 ; 5] (compare to LOI_{LIM} in SUBMOD).

The E–model has now become the major tool used by transmission planners for predicting the expected quality of future network scenarios. The multitude of different types of impairment introduced by modern networks, as well as the necessity to interconnect sections operated by different (private) companies, does not allow fixed limits for single parameters to be given as it was the case several years ago. To assure that the predictions are reasonably precise, it is important to know the range of validity of such a model. This will be investigated in more detail in Chapter 7.

3. INSTRUMENTAL MODELS FOR SINGLE TRANSMISSION ASPECTS

As stated above, a second class of instrumental prediction models exists which does not aim to predict the overall quality mouth–to–ear, but to estimate the quality degradation due to a certain piece of equipment, e.g. a codec. The measures are designed to estimate the mean opinion obtained in a listening–only test, so some characteristics of a natural telephone communication are included, namely the sending frequency response of the telephone (normally modeled by the IRS characteristic), and sometimes room noise at the send side. It should be emphasized that the target quantity to be estimated is the overall opinion determined in a limited set of listening–only tests, namely Y_{LQ} or Y_{LE}. So it can be assumed that quality elements which are not included in the judgment will not be modeled. The network planning models discussed above also make predictions for MOS, but they are based on many different sources of information, including conversation tests, user surveys, network experts' experience, etc. It can be expected that the broader basis helps to make them more robust in their prediction validity (perhaps less in accuracy).

The general structure of many of the instrumental models follows one of two types, as discussed by Hauenstein (1997). These structures are depicted in

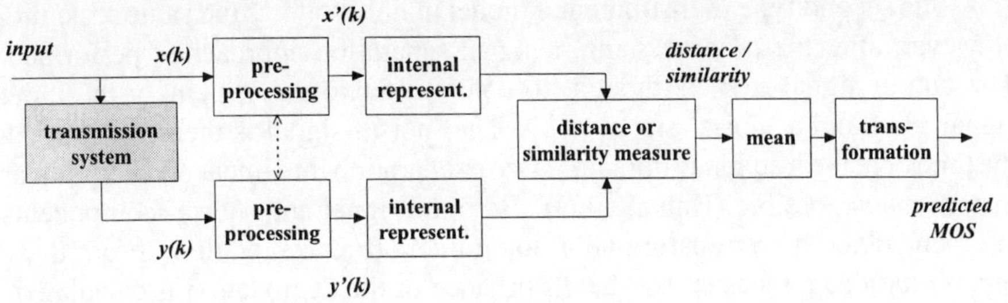

Figure 5.1. General structure of an instrumental model for single transmission aspects, type 1 (according to Hauenstein (1997)).

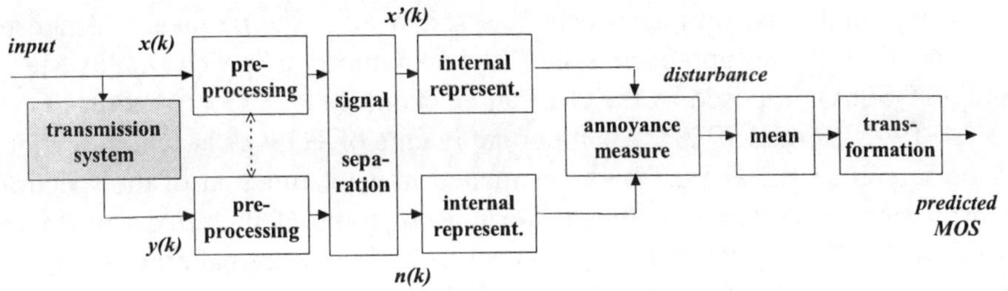

Figure 5.2. General structure of an instrumental model for single transmission aspects, typc 2 (according to Hauenstein (1997)).

Figures 5.1 and 5.2. The input and the output signals of the transmission system under test are subject to a preprocessing. This mainly serves to compensate for some of the features in the transmitted signal which do not directly lead to quality degradations, e.g. delay and overall attenuation. Sometimes a correction for linear frequency distortions and a speech activity detection are included as well. In the type 1 structure both preprocessed signals are then subject to a transformation onto an internal representation. An internal representation used very often is one which is more or less oriented towards the human auditory system. Hauenstein (1997) discusses a number of different psychoacoustically motivated loudness representations. Between the internal representations of the input and output signal a distance or similarity measure is then calculated. Also in this point, a variety of different approaches has been used. All the described steps are repeated for every segment of the speech signal. Then a mean is calculated across all parts of the signal, sometimes taking into account threshold effects or time–scale remember effects. The distance or similarity calculated in this way is finally transformed into an estimation or prediction of an overall quality judgment, e.g. Y_{LQ} or Y_{LE}. This transformation can be done e.g. by an $arc\ tan$–function.

The second type of instrumental model mainly follows the same structure. However, after the preprocessing, a signal separation approach is performed. The output signal $y(k)$ is thought to consist of two parts, namely the input signal $x(k)$ and a 'noise' signal $n(k)$. The 'noise' signal is then estimated in the time or in the frequency domain. Also a separation into linear and non–linear distortions is possible (Halka, 1993). Both the signal and 'noise' components are then subject to a transformation to an internal representation. From these representations a measure for the disturbance of the degradation is calculated.

The measure now standardized by the ITU–T, called 'Perceptual Speech Quality Measure', PSQM (ITU–T Rec. P.861, 1996), follows, in principle, the type 1 structure. The internal representation consists of a modified loudness calculation called the "compressed loudness", and the method takes into account the background noise during speech pauses, as well as the frequency response of the handset. A new approach is the 'Telekom Objective Speech Quality Measure' (TOSQA) proposed by the German Telekom (ITU–T Contribution COM 12–34, 1997; Berger, 1998). An important feature of TOSQA is a modification of the reference signal $x(k)$ by an estimated transfer function of the spectral transmission characteristics, thus balancing out some of the effects of linear distortions. The internal representation consists of loudness patterns which are calculated in variable frequency bands. An MOS prediction is calculated from the similarity between input and output representations.

Several other measures have been proposed. The better ones yield a high correlation with auditory listening–test results around 0.95 for many test conditions. However, the scope of validity of such measures is still not fixed. It should be noted that it is not too difficult to develop a model which shows a high degree of correlation with a known data set, provided that a sufficiently large number of free parameters can be chosen. The difficulty lies in developing a model which performs well even with unknown data, using a limited set of free parameters. It is expected that newer measures will additionally try to include effects of room noise and bit errors. The modeling of frequency distortions may be of some interest for a future development of the E-model. This will be discussed in Section 4 of Chapter 7.

4. CALL SET–UP AND CALL COMPLETION MODELS

Delays or the impossibility of setting up and terminating a call rank below the impairment of an actual transmission in terms of user dissatisfaction. This is especially the case for the conventional wirebound network, both in analog or digital configurations. Thus, there was no urgent need for developing models which try to predict the effects of the network access, call set–up or call termination phase. The situation is now different for mobile telecommunication networks, and for networks which make use of the internet. Although the user expectation plays a major role in the individual opinion of what is acceptable for a certain type of network, models developed for traditional wirebound networks can give a rough indication of to what extent user satisfaction is disturbed by an imperfect set–up and termination of a call.

Kort (1983) describes a model which has been developed at AT&T for their PSTN. It is based on subjective testing in a controlled laboratory environment, using 63 test subjects who performed conversation experiments. Similar experiments have not been replicated, so there are no complementary data to confirm or reject the model's predictions. The so–called "call completion model" combines the effects of an imperfect transmission with the difficulties in successfully setting up the call. It has two underlying models, the transmission rating model BcTR (which was described in Section 2) and a call set–up model. Both models result in a rating score R_{LNE} and R_{CS}, respectively, which are combined to form a prediction for call completion opinion R_{CC} in the following way:

$$R_{CC} = R_{CS}^{\alpha} \cdot R_{LNE}^{(1-\alpha)} \tag{5.20}$$

where

$$\alpha = \frac{1}{1 + 21.13\, log(1 + t_{call}/578.3)} \tag{5.21}$$

and t_{call} is the conversation time of the call (in s).

The call set–up model itself identifies the following deficiencies in the network access and set–up phase:

$$R_{CS} = 85.97 \cdot e^{[-\frac{t_{na}}{66.64} - \frac{t_{pd}}{59.95} - \frac{N_{DDI}}{104.7} - \frac{t_{WTI}}{192.7}]} \tag{5.22}$$

here

t_{na} is the dial–tone delay of the successful attempt (in s),

t_{pd} is the post–dialing delay of the successful attempt (in s),

N_{DDI} is the total number of digits dialed in all the ineffective attempts before the successful call, and

t_{WTI} is the total waiting time of all the ineffective attempts (in s).

The overall call completion rating, R_{CC}, can be translated into customer opinion in terms of the percentage of customers who would rate the completed call in each of the five categories of the MOS scale. The formulae given above can be used, e.g., to calculate a prediction of %GoB, assuming a certain level of transmission quality in terms of R_{LNE}. Some examples are given in Kort (1983).

A characteristic of the Bellcore call completion model is that R_{CC} largely depends on the conversation duration, in a monotonically falling relation. The shorter the call duration is, the more important are degradations due to delays in network access (dial tone delay) or call set–up (post–dialling delay) and due to ineffective attempts. This is true for longer delays or lower degradation in transmission. For more severe transmission impairments (lower R_{LNE}), however, the model predicts a contrasting behavior regarding slight call set–up degradation. The effect of the call set–up problems is very serious especially for conversations lasting less than 3 min. For longer conversation times, even major delays in call set–up are not predicted to have a strong influence. This can be explained by the user's memory of negative events. The user's opinion will be less negative if a call of sufficient duration follows a longer waiting time for it to be set–up.

For the quality of mobile GSM speech services, Verschoor (1994) gives a simple model aiming at comparison and evaluation. The starting points are two of the factors which are used in ITU–T Rec. E.800 (1994) to establish the definition of QoS, namely serveability and service integrity. Network performance quality elements are defined for all the phases of a mobile connection, i.e. network access phase, call set–up phase, execution phase and call termination phase:

- Network access phase: P_{naf} is the probability of failing or long lasting network access, when $t_{na} > t_{na,max}$.
- Call set–up phase: P_{suf} is the probability of failing or long lasting call set–up, when $t_{su} > t_{su,max}$.
- Execution phase: P_{cf} is the probability of connection failure, and G_{si} is the grade of service integrity, defined by

$$G_{si} = prob\left(\frac{t_{lq}}{t_{call}} > \alpha_{lq} \vee N_{HO} > f_{HO,max} \cdot t_{call}\right) \tag{5.23}$$

- Call termination phase: P_{ctf} is the probability of failing or long lasting call termination, when $t_{ct} > t_{ct,max}$.

In these expressions,

t_{na} is the time to reach network access,

t_{su} is the time required for call set–up,

t_{lq} is the summation of periods of insufficient low speech quality,

t_{ct} is the time required for call termination,

t_{call} is the call duration,

N_{HO} is the number of handovers during the call,

α_{lq} is the maximum allowable fraction of call duration during which speech quality is insufficient, and

$f_{HO,max}$ is the maximum allowable frequency of handovers.

In order to obtain one universal quality parameter for GSM speech services, the network quality parameters are combined to form the so–called Quality of Service Index I_{QoS} by means of weighting factors w in the following way:

$$I_{QoS} = w_{si}G_{si} + w_{naf}P_{naf} + w_{suf}P_{suf} + w_{cf}P_{cf} + w_{ctf}P_{ctf} \quad (5.24)$$

Verschoor (1994) gives indicative values for the parameters and weighting factors. The weighting factors are particularly interesting from a user quality point of view. The values are reproduced in Table 5.3. It becomes obvious that the relative importance attributed to the call execution phase is about eight times greater than that for all the other phases put together. Remarkably the access and set–up is not regarded to be of great importance. Connection failure has about the same weight as service integrity, i.e. the quality of the signal transmission during the call.

Table 5.3. Indicative values for the weights of the GSM Quality of Service Index (Verschoor, 1994).

call phase	weight	value
network access	w_{naf}	2
set–up	w_{suf}	2
execution	w_{cf}	20
	w_{si}	20
termination	w_{ctf}	1

It should be noted that the correctness of the model has not been shown. Moreover, there is no relation given between the Quality of Service Index I_{QoS} and any expression of user opinion. The QoS index varies between 0 (excellent quality) and 20 (very bad quality). Verschoor underlines that the quality predictions are for network planning purposes, more precisely for selecting base station parameters in order to reach optimum quality with an existing network of base stations. In conjunction with a traffic and user behavior model (which gives values for geographic behavior and call behavior of subscribers) the Quality of Service Index can be used to optimize e.g. a handover margin as a compromise between a too high number of handovers on one hand, and bad periods or connection failures on the other. 'Absolute' quality predictions which could serve as an indicator of overall network quality are not the aim of this model. The weighting factors may be decided upon by the network operator or service provider, according to his experience. They are a compromise between the quality experienced by the user and the cost of network operation.

5. CUSTOMER BEHAVIOR MODELS

This term refers to the observation of user reactions to strong degradations in the transmission. It covers abandonment or retrial of the call, or complaints to the network operator. So far, no models for user behavior in specific situations or under certain transmission path characteristics have been developed, e.g. regarding the adaptation of speaking style to quiet or noisy connections, due to too loud sidetone or room noise, or the user interaction in long delay conditions. Some of these effects are covered (though not explicitly modeled) by the transmission planning models, as long as their quality predictions agree with the results of conversation tests under natural conditions.

If the call initiation or the transmission performance is severely impaired, the customer, i.e. the user of the service, will either abandon the connection or the unsuccessful attempt, or he/she may complain to the operator, network operator or service provider. The latter behaviour generally only occurs in very rare cases; however, network operators and service providers pay attention to such behavior because it may give the company a bad image, and it may show impairments in the network which will affect a larger number of users.

Call attempts may be abandoned in any of the network access, call set–up and the transmission phases. Kort (1983) gives several formulae for the probability of abandoning before the dial tone, while dialing (depending on the number of digits to be dialed), and before the network response. These formulae are the result of observations and subjective experiments carried out on the specific AT&T PSTN. They are highly dependent on the actual network and on the experience of the customers. As an example, if the network response times improve, customers will be less inclined to wait before abandoning a call. Both response times and abandon times are subject to the network development. Therefore, the exact values calculated with these models are of very limited use for predicting customer behavior for abandonment.

For a given network response other than completion, customers will probably attempt a retrial. This probability depends on the network response (ring with no answer, busy, no such number, intercept, etc.), and will generally be in the area of 60–90 %. During an established connection, severe transmission impairments can lead to the user abandoning the call, and/or to complain to the operator or to the service provider. Based on customer call–back interviews and the experience with their own operated network, the probabilities for terminating a call early (%TME), for retrial and for operator complaints were modeled (Kort, 1983). The transmission impairment is described by the transmission rating R_{LNE} of the BcTR model. The resulting formulae are very similar in their structure:

$$\%\text{TME} = 1 - E\left(\frac{R_{LNE} - 36.30}{15.85}\right) \cdot [100\%] \tag{5.25}$$

$$\%\text{Retrial} = 1 - E\left(\frac{R_{LNE} - 29.50}{15.85}\right) \cdot [100\%] \tag{5.26}$$

$$\%\text{Operator Complaints} = 1 - E\left(\frac{R_{LNE} - 23.69}{15.85}\right) \cdot [100\%] \tag{5.27}$$

where $E(x)$ is the Gaussian error function defined in Equation G.27. The predictions are given in Figure 5.3. As the E–model transmission rating factor R is closely linked to R_{LNE}, the same behavior can be calculated with the E–model output. However, it should be noted that these functions largely reflect the situation of North–American wirebound networks during the early 1980s. They are not applicable as they stand to digital networks (allowing faster dialing and special dialing features) and especially not to mobile networks. It will be interesting to develop similar models for telephony over internet protocol (IP). In this case a major influencing factor will be the users' expectation of the computer response, not just of the network response.

6. APPLICATION AND CLASSIFICATION OF PREDICTION MODELS

Different models for predicting quality in telecommunication networks have been presented so far. One question that arises is how to apply such models in order to obtain a valuable estimation of specific network parts or of the network as a whole.

The most common way of using models is to perform comparative calculations on different network scenarios. These scenarios may already be established or due to be set–up in the future. The model predictions can help to choose one of different network element possibilities. In a more absolute way, model predictions are also used to give direct estimates of expected integral quality of a certain network configuration. In this regard, the model prediction is used by the network operator, installator or service provider to prove (in front of the customer, e.g. the user of a private network) that his installation complies

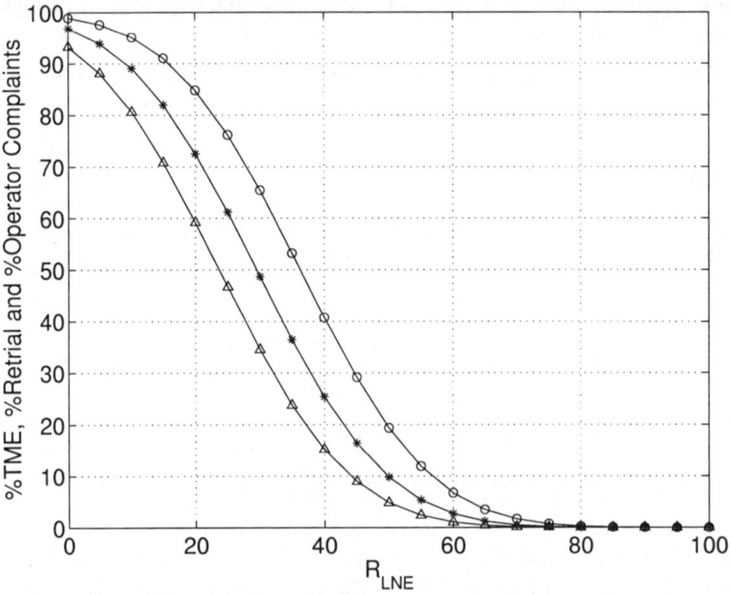

Figure 5.3. User behavior as a function of the BcTR transmission rating R_{LNE}. o: %TME; *: %Retrial; Δ: %Operator Complaints.

with recognized rules of quality network planning. The prediction is also used to prove expected network quality against potential rival companies. Especially in the latter case model predictions may be the result of target–directed network tuning, in the sense that equipment is used whose model predictions do not lead to realistic, but rather optimistic user opinion estimations.

If many or (better) all of the parameters of a connection are known or can be measured, a quality estimation of an imaginary "mean user" can be calculated. It should be noted, however, that this estimate will not correspond to the quality event a single user has. The measurement of input parameters can be done in either an intrusive or non–intrusive way, as discussed in Section 5 of Chapter 7. For parameters which cannot be measured, realistic or "typical" values will be assumed. If the quality predictions and the associated measurement cover a larger period of time and several parts of the network, they can be used to identify network problems and thus help to maintain and evolve networks. The transformation of physically measured parameters to prediction models is one major point of discussion in this domain.

One final use of prediction models is to generate input values using Monte–Carlo simulation models. The underlying assumption is that network impairments are characterized by statistical distributions, and that the effect of the accumulation of various impairments may be calculated in the combined way the prediction models offer. Kort (1983) describes such a model. The Monte-Carlo simulation generates a random sequence of telephone calls including disposition and associated waiting times. Transmission performance, call set–up and the combined call completion performance are calculated according to the models described above. The simulation output comprises the predictions of transmission performance, call set–up performance and call completion performance, as well as statistics of ineffective calls. Separate runs of the simulation can be performed on different parts of the network, and for different mileage, etc.

When interpreting and applying the model outputs, it is important to note which quality features the prediction models refer to. All of the instrumental methods discussed in Section 3 as well as all the articulation measures (which were not discussed here) only relate to one–way voice transmission quality, because these measures have been established to predict a quality rating obtained in a listening–only test situation. Also transmission planning models, like the E–model or SUBMOD, predict many features relating to one–way voice transmission. On the other hand, the predictions for 'delayed impairments' (talker echo, absolute delay) refer to the communication efficiency. It will be shown in Chapter 7 that model predictions for some of these features are not valid for general situations. The E–model expectation factor primarily relates to some features concerning the comfort of a service (advantage of access). However, it cannot be denied that the costs have an influence as well. Models relating to other comfort features are the call set–up and the call completion models discussed in Section 4. Customer behavior models aim at predicting user behavior in terms of TME, retrial and operator complaints. Thus, they provide estimates of acceptability, but only for very low quality services. Positive acceptability models may exist in the marketing area, but they will not be based on characteristics of the transmission configuration.

7. SUMMARY

In the planning of telecommunication networks, it is desirable to have estimations of the quality to be experienced by a potential user even before the respective network has been set up. This permits the design of high quality systems and services satisfying the users' needs. Based on auditory test data collected over the years and on experiences with existing networks, transmission planning models were developed. These models calculate quality predictions either from instrumentally measured or reference data of the telephone connection of interest (transmission planning models) or from the input and output signals of a defined part of the transmission chain (instrumental models for single transmission aspects). Several transmission planning models were developed by different telecommunication administrations, taking into account

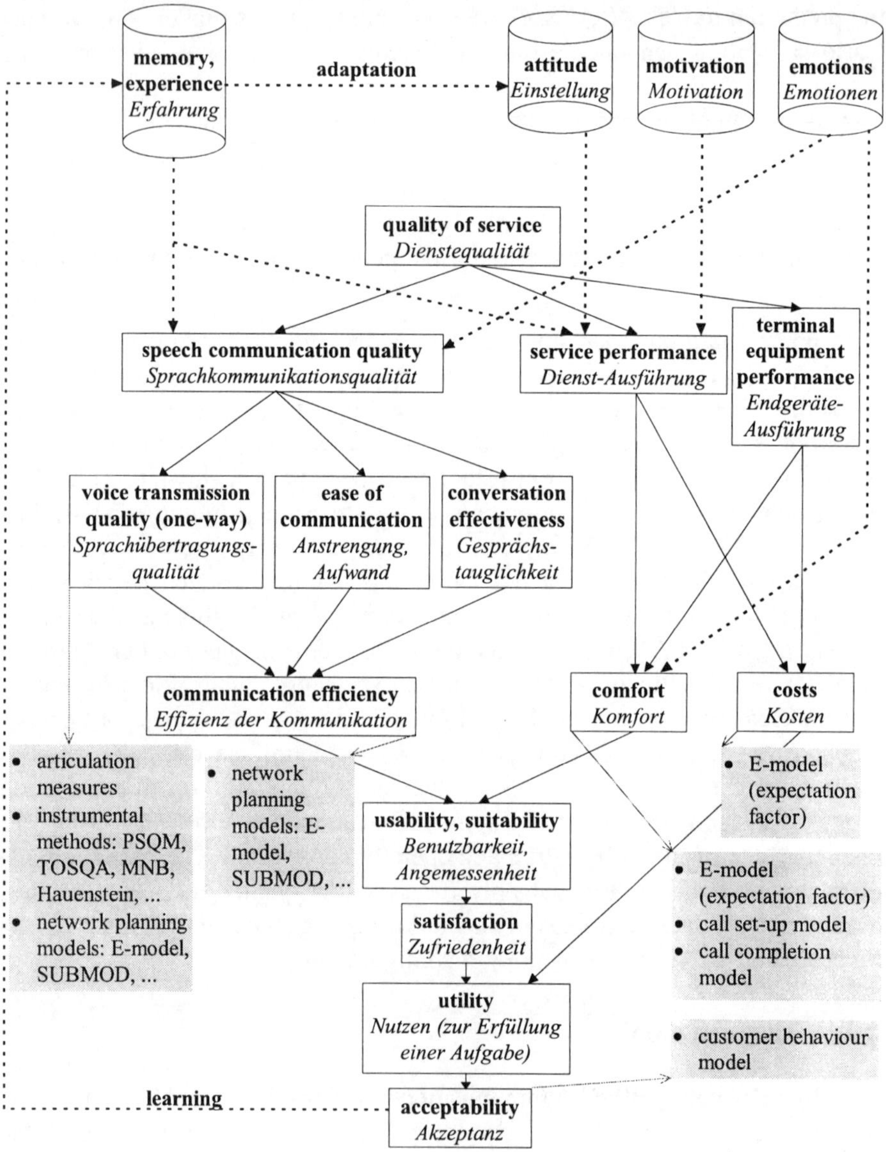

Figure 5.4. Classification of prediction models in the QoS schematic. German terms are given in italics.

most of the degradations occurring in traditional networks. Two members of different classes of models are analyzed here in detail: one making use of frequency–dependent quantities (SUBMOD) and one using scalar input parameters (the E–model). The E–model is now the only model recommended for use in transmission planning by the ITU–T.

The SUBMOD model is based on the loudness rating model described in Section 1.1 of Chapter 3. Instead of loudness, a quantity called 'listening–opinion index' (LOI) is calculated in a first step. This is performed by integrating over the contributions in terms of sensation levels. The LOI can then be transformed into estimates of listening–effort MOS and later conversation MOS. The latter transformation includes the sidetone path characteristics.

Two representatives of the scalar parameter models are the Bellcore transmission rating model (BcTR) and the E–model. They make use of an internal 'transmission rating scale', which should render the quality predictions independent of individual tests. Nevertheless, a transformation is given for calculating conversational MOS from the transmission rating R. It is shown that the transmission rating scale is used in a similar way both in the BcTR and in the E–model. The basic assumption of the E–model is that the contribution of different psychological factors to the overall quality is additive on the transmission rating scale. Thus, different types of impairment can be handled, by simply calculating individual so–called 'impairment factors' for each impairment and subsequently adding the numerical values. The sources of this assumption are experiments carried out by Allnatt and coworkers on subjective quality ratings of impaired video pictures. The psychophysical bases are discussed in detail, though a verification of the principle is still outstanding. The analysis of the quality of E–model predictions in Chapter 7 will furnish experimental data on the validity of this principle, taking several combinations of impairments into account. The E–model calculates four different impairment factors, one relating to the basic signal–to–noise ratio of the connection, a second representing impairments occurring simultaneous to the speech signal (e.g. a too loud connection, non–optimum talker sidetone and quantizing distortion), a third for impairments occurring delayed with respect to the speech signal (talker and listener echo, overall delay), and a forth capturing impairments caused by more complicated new equipment. For reasons of simplicity of the model, the latter 'equipment impairment factor' which was originally designed to represent the effects of low–bitrate codecs on voice–transmission quality now is being used for a couple of different types of equipment. A simplified correction for user expectation differing from the standard is also included in the model by means of a so–called 'expectation factor' A. This factor should represent the advantage of access the user of a special system (e.g. a mobile phone) experiences over the conventional wirebound service. The validity of the expectation factor will be discussed in more detail in Section 4 of Chapter 6.

Apart from the transmission rating models, instrumental models exist aiming at predicting quality degradations for single aspects of the transmission chain. The major application is the prediction of effects due to low–bitrate codecs, but extensions are planned in order to produce valid predictions in the presence of other impairments. The general structure of such instrumental models is presented. In principle, a distance or similarity measure is calculated from the input and the output signal of a transmission system. Several models make use of psychoacoustic characteristics of human hearing in order to produce perceptively relevant measures also for transmission systems which were not used for the establishment of the model. The distance or similarity measures are then transformed to predict quality estimates as they would be obtained in a listening–only test situation. Due to the inherent structure of the models, they only aim at predicting one–way voice transmission quality for a specific part of the transmission system. In contrast, network planning models take the whole transmission mouth–to–ear into account, including effects occurring in conversation.

The models described so far relate to the speech communication quality provided by a system or service. On the other side, service performance resulting in comfort for the user depends on additional factors in the network access phase, the call set–up phase and the call termination phase. Two different approaches are presented which refer to the expenses required from the user, e.g. in terms of time and dialing effort. In the call execution phase, service integrity is of major importance for the usability. Services showing low call integrity will result in strong user reactions such as abandonment of the call, retrial, or complaints to the network operator or service provider. Simple modeling approaches for this behavior are presented. The validity especially of service performance related models will be very limited to the system they were designed for. Nevertheless, they will reach new importance in the future, because new types of services first have to find a balance between service performance and economic factors (number of potential users, costs, etc.).

The presentation of modeling approaches shows that different features of quality experienced by the user are addressed. The models' input parameters, on the other hand, are quality elements which are (to a large extent) in the hands of the network planner. The models thus provide a link between the quality elements and the quality features. A detailed investigation in Chapters 6 and 7 will show the validity of performing such a link. Especially by investigating user expectation and conversational influences the limitations of model predictions become clear.

Chapter 6

RELATIONS BETWEEN FACTORS GOVERNING THE QUALITY OF SERVICE

The desires of transmission planners are high. It is worth citing Richards (1973, p. 125) to describe the dilemma:

> "It would be convenient if the telephone transmission performance of a given speech link could be expressed as a unique value along a unidimensional scale graduated in clearly definable units. Certain rather simple relevant subjective criteria do appear to be unidimensional, such as 'comparative loudness' and 'comparative fitness for a defined purpose' and much advantage can be taken of these properties. Unfortunately, however, these only enable a particular set of speech links to be ranked according to certain criteria and they do not yield an unequivocal absolute scale which would have universal meaning."

In practical experiments, the experimental and environmental conditions have a strong influence on the quality rating. Nevertheless many experiments yield a score which can be fairly well reproduced under similar test conditions. Although certain characteristics regarded as being unidimensional can be rated on single unidimensional scales, scaling mainly results in a ranking of the tested conditions with regard to the scale dimension. Such ratings cannot generally be combined to form a quality prediction on an absolute universal scale. In this chapter some of the relations between factors contributing to the quality of a service are highlighted. For this purpose, it is worth investigating first the factors governing quality assessment experiments (Sections 1 to 3), and then more general QoS influencing factors such as expectation and costs. The discussion results in an approach for the assessment of a unidimensional quality feature similar to the 'transmission rating' of the E–model. The collection of influencing factors is not complete, but seems to cover several points which have an important effect on quality perception. They are therefore worth considering in the planning of telecommunication networks as well.

1. 'RELATIVE' QUALITY VERSUS 'ABSOLUTE' QUALITY

Earlier it was discussed that quality is not a fixed entity which is inseparably linked to a perceptive event (e.g. telephone speech). In contrast, quality is an event itself, because it is attributed to the perceptive event by the conversing and judging person. The relation is established instantaneously and depends on the judging person, his/her experience and expectations, the situation, motivation for the call, etc. Because of this, a quality judgment can only be relative to the situation being assessed, and will never be 'absolute' in the sense of independence from the assessment situation. This fact has to be taken into account when quality assessment is planned, or the results of a quality assessment process are interpreted. In this section, the influence of the quality reference in auditory tests is discussed in more detail. A specially designed conversation test (SSRCT) is carried out in order to quantify the influence. Methods aiming at neutralizing the reference for standard opinion test results will be analyzed in Section 2.

Especially in absolute category rating tests, the most common method for assessing the quality of telephone transmission systems, context effects occur which have a strong influence on the test subject's rating. In a listening–only test, the subjects are asked to listen to a relatively large number of test conditions (usually speech samples) and to make a judgment on a quality–related feature for each stimulus or stimulus pair. Though the subject is instructed to make a rating based only on the actual stimulus, there are several influences on the judgment which do not depend on the stimulus to be rated. According to (Duncanson, date unknown), the effects can be categorized in the following way:

- Distribution effects: The previous experience with the test object as well as the range of conditions and their distribution within the test have a major effect on the results. Test subjects tend to shift their standard of judgment (which is based on previous experience) towards the set of objects to be rated. This results in more critical judgments when the quality range within the test is high, and more optimistic ratings when the quality range of test objects is rather low. Besides the condition range, also the shape of the distribution of test objects is important. Parducci (1963) summarizes the results of several psychophysical experiments with category judgment tasks on different modalities into two major principles: Test subjects tend to divide the range of test objects into subranges for each category, and they tend to use the categories equally often. Therefore, if the test objects are distributed with a normal distribution, there is a tendency for the subjects to make the center categories narrower and the end categories wider, so that the judgments falling into each category remain the same with respect to a uniform distribution. In order to obtain veridical level information in category rating experiments, the stimulus range should be 'ecologically valid', i.e. the intensity range of the stimuli should represent the intensity range experienced in everyday situations outside the laboratory (Müller et al., 1996). Additionally, the scale should be 'psychologically valid', with verbal descriptions as they are commonly used in everyday life.

- Order effects: The judgment of the current test object is also influenced by the preceding test objects. In general, judgment is displaced towards the rating of the last preceding test object. If the average value of the test objects shifts either up or down, this shift is exaggerated in the judgments immediately following the shift. The influence results from a number of preceding test objects, but the influence of the most recent test object is the strongest.

- Anchor effects: Besides the experiences immediately preceding the test object, the judging subject makes an indirect and unconscious comparison with some kind of reference. This reference may be the mean overall experience the test subject has had with the test object (e.g. a telephone connection) in the past, or it may be given as an explicit reference before or within the experiment. In general, a reference which lies outside the 'normal' experience provokes a shift in judgment away from the anchor. If the anchor lies within the range of 'normal' experiences and is specifically pointed out to the test subjects, it can serve to reduce variability in the judgments around the value of the anchor. Heller (1990) states that a quantitative judgment obtained by means of category scaling is only valid if it is clear by what the subject has oriented himself/herself. In an experiment where several test conditions are presented successively without the preceding presentation of an anchor, orientation takes place by means of the scale. The subject establishes a relationship between the first stimulus and the scale, and this relation determines the use of the scale for the rest of the experiment.

Especially in LOTs it is particularly important for the outcome of an experiment whether the test subjects are familiar with the range of test conditions using the presentation of anchors at the beginning, or whether they do not have anchors for orientation. The use of anchors allows the test subject to make better use of the whole category scale; however, in this way the range of test conditions is transformed more or less directly to the set of categories, irrespective of the category labels. Thus, the use of the categories is different for oriented (with anchor) and non–oriented category judgments. In conversation tests, the situation differs, because the more natural test situation suggests an indirect comparison to be made with the normal telephone connection experienced in everyday life. The comparison with long–term overall experience would be ideal for a quality assessment test. However, it can only be achieved by performing user surveys. The survey technique, on the other hand, does not allow complete control over the physical characteristics of the connection.

An interesting observation regarding the orientation even in user surveys was reported by Duncanson (1969). For a sample of normal users of the transatlantic telephone service, it was found that the rating of single, recent telephone calls leads to better results then the one of subjectively averaged telephone calls of the same type made some time ago. One reasonable explanation is that a subject remembers very bad past conditions more readily than other conditions of good quality, and attaches higher significance to these conditions in the subjective averaging process. As a consequence, comparisons between ratings obtained in different instances (e.g. relating to a recently completed call and a usual, average call) are not valid.

Context effects occurring in conversation tests under laboratory conditions are addressed in the following by comparing standard SCT results to those obtained in an non–influenced way for only one test condition. A specially designed test has been performed recently in a way that each subject rated only one connection, without having been confronted with an anchor beforehand. A slightly modified version of the short conversation test scenarios was used to stimulate the conversation. Test subjects were asked to participate in the university area at random (however, the age group 20–35 years was over–represented). The range of test conditions included very bad conditions (MNRU with Q = 5 dB) up to clear connections using the default parameter settings of Tab. 3/ITU-T Rec. G.175. 110 test subjects took part in the test, so that 11 ratings could be collected for each of the 10 test conditions (Darmoe, 1998). The test conditions were taken from the conditions used in an ordinary conversation test with similar test instructions (SCT). In this way a comparison between the results of the rating of only one test condition (here called "single stimulus rating conversation test", SSRCT) and the ordinary conversation test (SCT) is possible. However, it should be kept in mind that the test subjects were different for both experiments.

The MOS results depicted in Figure 6.1 regarding the MNRU circuit conditions show that the usage of the full rating scale range is better for the SSRCT than for the SCT. In all, MOS ratings of the SSRCT lie in the range MOS $\in [1.36; 4.55]$, while for the SCT MOS $\in [1.79; 3.93]$. When subjects are asked to judge several conditions in a sequence, it seems that they avoid using the outer categories of the MOS scale in order to be free to use them for potentially better or worse circuit conditions presented later in the experiment. This was already discussed in Section 8.2 of Chapter 4, item 5. Especially in the middle of the MOS rating scale, the judgments obtained in the SSRCT are more than 0.5 MOS intervals better (except MNRU with $Q = 20$ dB) than those

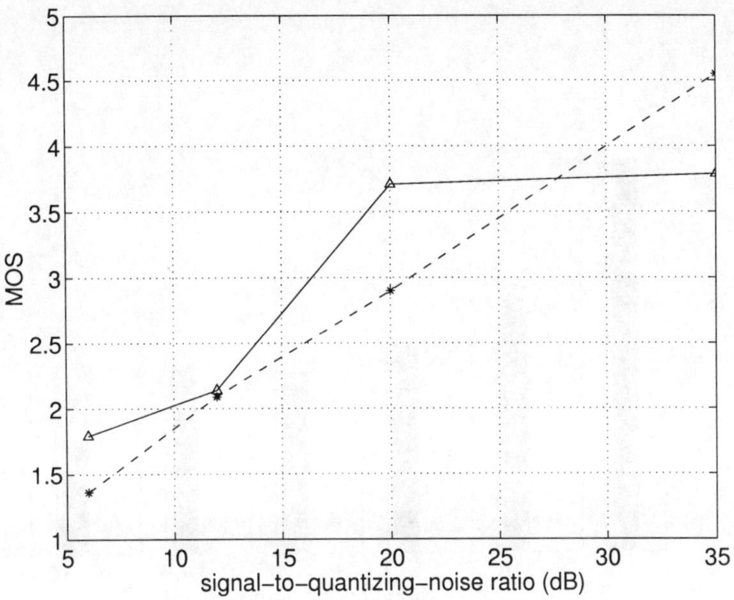

Figure 6.1. Comparison of SCT and SSRCT results for MNRU conditions. Δ: SCT results; *: SSRCT results.

obtained in the SCT, see Figure 6.2. It is possible that these test conditions correspond more to the verbal "fair" category (German label: *"ordentlich"*) within the average subjects' quality experience, but because better conditions were included in the condition set of the SCT, the ratings obtained there are lower. This would be an indication of a context effect of the anchor type. The connection corresponding to the everyday experience of the test subjects corresponds to a rating of MOS $\sim$ 4.0, see Figure 6.3. Thus the internal anchor representing usual telephone connections is labeled with the verbal attribute "good" (German label: *"gut"*). This global reference is not in the center of the rating scale, so the scale is used in an unbalanced way compared to the reference. Only two of the ten test conditions of the SSRCT are rated equivalent or better than the usual connection. The same holds true for the SCT conditions, where only two of 27 conditions obtain an equivalent integral quality rating.

The investigation presented here does not literally relate to the factors governing the quality of a service. Nonetheless, if communication efficiency is investigated, this will be done by performing auditory tests, which in turn are strongly influenced by the respective internal and external references. In summary, test results obtained using the SSRCT technique show that an influence of the condition range and the test situation exists in a naturally designed conversation test situation as well. Thus, ACR tests do not provide 'absolute'

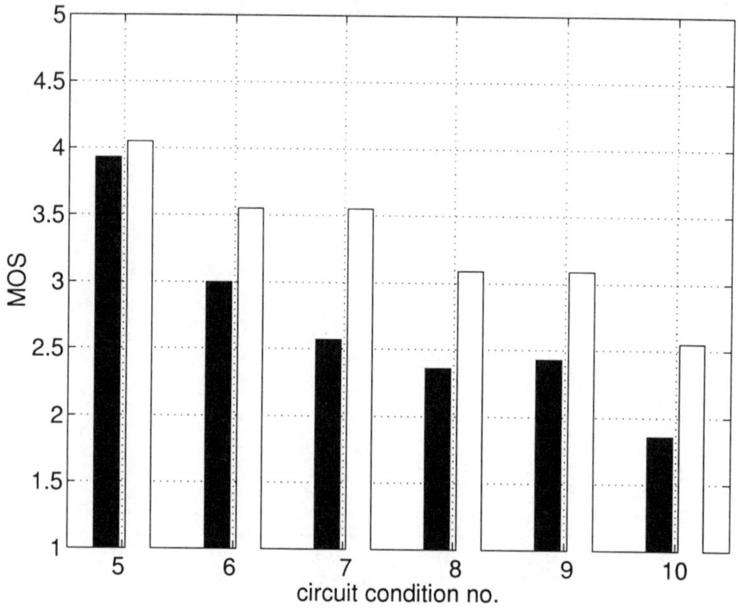

Figure 6.2. Comparison of SCT and SSRCT results for talker echo and codec conditions. Black: SCT results; white: SSRCT results.

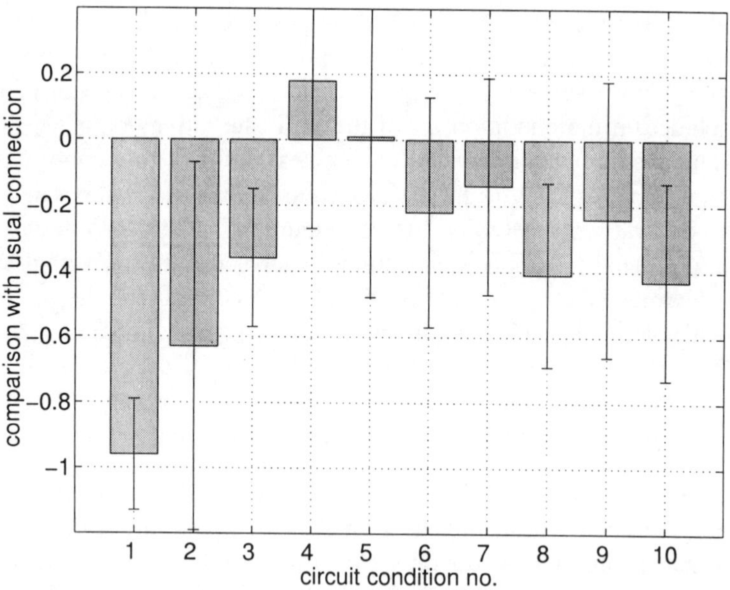

Figure 6.3. Rating of the SSRCT conditions in comparison to the subjects' own subscriber line at home (usual connection). '0': equal; '-1': much worse than usual; '1': much better than usual. Bars show mean ratings and standard deviations.

quality estimates, in the sense of referencing the usual situation in everyday life. It can be expected that the influence will be stronger in a listening–only situation, when many more test stimuli are presented in a shorter period of time. The test also reveals that the MOS scale is not appropriately designed to have an image of the 'normal' situation in the center of the scale. Instead, the lower quality area (below the standard telephone quality) is more extended.

2. REFERENCES AND NORMALIZATION

If absolute quality ratings cannot be obtained, it is often preferable to have a relative rating in the form of a comparison with an adjustable reference condition. In some senses, the loudness ratings and the original concept of 'impairment' already adhere to this idea. The loudness rating can be obtained, other than by calculation from the sensitivity characteristics of the path, by performing listening–only experiments as the equivalent attenuator setting of a reference system for obtaining the same impression of loudness. The concept of 'impairment', in its original definition, is the equivalent attenuator setting of a reference system for obtaining the same impression of overall quality.

The reference system now used and standardized for telephony is the so–called modulated noise reference unit, MNRU. It is standardized in a narrow–band and a wide–band version, for analog as well as digital realizations (ITU–T Rec. P.810, 1996). The principle is very simple, see Figure 6.4. All the DC components are removed from the input signal. The signal is distributed to two paths, the signal path and the noise path. In the signal path, the speech signal remains unchanged with the exception of linear amplification or attenuation. In the noise path, the speech signal is multiplied by the white wideband noise. The resulting signal–correlated noise is amplified or attenuated and then signal and noise are added. The resulting degraded signal is finally band–filtered to the frequency range 100–3400 Hz. The two attenuators allow a fixed signal–to–noise ratio Q of the signal degraded by speech–correlated noise to be specified. The MNRU is recommended for use as a reference when evaluating the subjective performance of digital speech transmission systems. It has been widely used for transforming listening–only test results for speech codecs at medium and low bitrates. The degradation introduced is very similar in nature to the quantizing distortion caused by logarithmic PCM coding, however of freely adjustable amount.

Apart from the MNRU other reference systems have been proposed. One very simple method is to use uncorrelated white noise. Rothauser et al. (1968) report a test where uncorrelated as well as signal–correlated noise were used as a reference. For tests involving LP, HP and vocoder degradations, both references produce well correlated results in the rank orders of the test conditions.

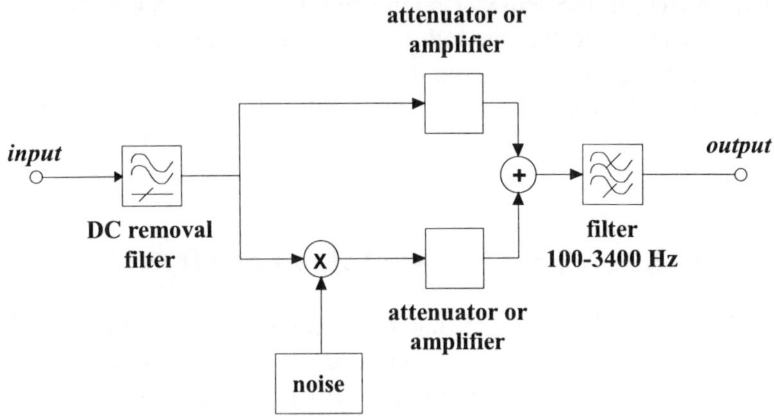

Figure 6.4. Schematic of a narrow–band MNRU according to ITU–T Rec. P.810 (1996).

A different reference system which has been proposed for text–to–speech synthesis is the "time–and–frequency warping" algorithm, TFW (Johnston, 1997). The principle is very simple: The speech signal is recorded at a constant sample frequency and then played back at a variable rate. Given that the ith sample is taken at time $T_i = \frac{i}{F_s}$, it is played back at time

$$T_i' = \frac{1}{F_s + A_w\ sin(\frac{2\pi i F_w}{F_s})} \tag{6.1}$$

where A_w and F_w are the amplitude and the frequency of the warping. The warping frequency is fixed at 6.67 Hz so that the amplitude of the warping is the only variable parameter of the reference. The sound degradation generated by TFW is a controlled "wow" and "flutter", which is similar in nature to the sound of traditional text–to–speech synthesizers. It has been used for TTS as well as automatic speech recognition system evaluation (Johnston, 1996). Further reference degradations are the "temporally correlated noise" developed by Bellcore (Cotton, 1992), consisting in a controlled addition and deletion of speech signal samples, as well as the so–called "S–reference" from NTT (Asanuma and Nagabuchi, 1994) which makes use of the source–filter model commonly used in speech processing.

The question arises as to whether a reference system should be similar in nature to the degradations of the test signal, or whether it is more convenient to use various and differently sounding references. Hecker and Williams (1966) report a study of preference testing methods using different reference signals. In a paired comparison test, the system under test was a) compared to several levels of degradation (SNR) of one reference system, and b) compared to several distinct types of reference systems, pre–adjusted in such a way that each was reliably preferred to the next in order. The preference judgments were then converted to a one–dimensional scale on which the system under test is located in relation to the reference systems. It is argued that method a) is insufficiently precise because the listener could be diverted from his/her task of making preference judgments to a task of matching the test system with one of several reference conditions. The experimental result is a larger deviation of the preference scores. However, it can be expected that the choice of reference conditions in method b) will have a great influence on the test outcome. Especially when different types of impairment are seen in relation to each other, the preference judgments obtained will be of a multidimensional nature, thus not justifying the establishment of a one–dimensional preference scale for comparison with the system under test. In contrast, data collection should reflect the nature of the judgments. A reasonable procedure would then be to make a multidimensional analysis of the preference judgment, as demonstrated by McDermott (1969). Rothauser et al. (1968) also express the desire to use reference signals which are similar in nature to the anticipated test signals, in order to improve the reliability of the subjective comparison.

Besides establishing rank orders with systems under test, reference systems are also used to transform test results to a more general level. For telephony and the MNRU, test results obtained in different experiments are recommended to be transformed by means of the so–called "equivalent Q method" (ITU–T Rec. G.113, 1996). According to this method, there is a fixed relationship between MOS ratings obtained for an MNRU and the corresponding signal–to–noise ratio Q of the MNRU. In ITU–T Rec. G.113 (1996) a $tanh$–function is assumed:

$$MOS = 1 + A + B \cdot tanh\left(\frac{Q - Q_m}{C}\right) \tag{6.2}$$

where A, B, C and Q_m are constants obtained by curve–fitting for the set of MOS(Q) pairs obtained in the specific test. After determining these constants, an MOS value for e.g. a specific codec can be transformed into an "equivalent" Q:

$$Q = Qm + \frac{C}{2} \cdot ln\,\frac{B - A - 1 + MOS}{B + A + 1 - MOS} \tag{6.3}$$

The "equivalent" value for Q can then be used to calculate an overall transmission rating and a transformed value for MOS, using the necessary formulae of the E–model, see Section 2.2 of Chapter 5, and Appendix G, Formulae G.12 – G.14, G.7 and 5.19.

The use of a reference system relies on the assumption of equivalence between a test condition and the reference degradation. In principle, two approaches are possible:

- Direct comparison of the digital system with the MNRU (PC, adjustment method).
- Indirect comparison and later calibration using the MNRU judgments as transformation anchors. This is the case for the "equivalent Q" method described above.

Gleiss (1987b) states that two requirements must be fulfilled for the MNRU:

- The judgment of the MNRU conditions should be stable on an absolute quality scale. Especially, it should not be correlated to the other test conditions.
- The sound quality of the MNRU should be similar to the circuits under test.

According to Gleiss and to own experiences, only little experimental evidence exists to show that these two requirements can be met. Especially for modern speech processing equipment (low–bitrate codecs, artifacts originating from echo cancellers, clipping, etc.) the latter requirement will very often not be fulfilled. Nevertheless, in the case of the "equivalent Q" method, it is interesting to see whether the equivalent setting of an MNRU (indirect comparison) really corresponds to the one which would be rated as an equivalent in an isopreference test (direct comparison). In the frame of this book, this has been done for several test conditions (including codecs, room noise at the send side, and frequency distortions) in listening–only and conversational tests (indirect comparison), and with two reference systems (MNRU and TFW) in a direct listening–only isopreference test. In the isopreference test, the subjects could directly compare the test sample to a fine–graded set of reference conditions of one reference type. They had to determine the reference setting which sounded as badly as the test signal. All the results were transformed into rank orders, see Figure 6.5. The condition parameters are given in Appendix H.1.

From the figure it seems obvious that, in general, there is a good agreement between the rank orders obtained in the experiments. However, for some test conditions differences exist not only between TFW and MNRU in the isopreference test (A & F) regarding the distribution on the scale and the order of conditions. Also between the test results obtained in different listening–only and conversation experiments (D-E) rank orders are sometimes reversed or have a different distance (e.g. conditions 13/14 in B and C, conditions 9/10 in D and E). This shows that there is no universal relationship between MOS results for MNRU and those obtained for test conditions that are different in perceived nature. A transformation using the MNRU (or any other reference which is not similar to the item under test) may therefore result in erroneous opinion scores.

As a further example, results of two tests carried out here and investigating talker echo (SCT) and low–bitrate codecs (LOT) are given. The results for talker

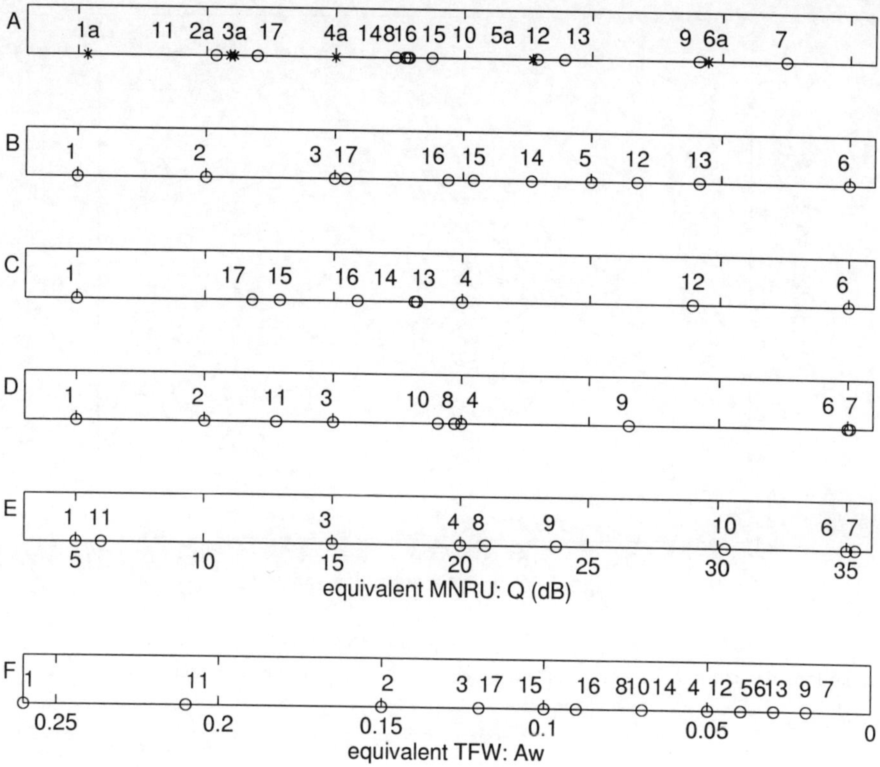

Figure 6.5. Rank order of test conditions in direct and indirect comparison experiments. A: MNRU, direct comparison; B: LOT, indirect comparison; C: SCT, indirect comparison; D: LOT, indirect comparison; E: SCT, indirect comparison; F: TFW, direct comparison. Condition numbers refer to Table H.1 of Appendix H.

echo have been transformed, first by means of the "equivalent Q" method, and then by a linear transformation using an echo curve with $T = 100$ ms one–way delay. The results for MNRU (codec and talker echo test) are given in Figure 6.6, the original results for talker echo in Figure 6.7. These results have been transformed according to both methods, see Figures 6.8 and 6.9. For comparison, the predictions of the E–model for talker echo are given as slashed lines. These predictions turned out to match well with several test results obtained in different laboratories, cf. Section 1 of Chapter 7. From Figure 6.8 it can be seen that the transformation using the MNRU leads to much more optimistic results, even higher than the E–model predictions. The linear transformation (Figure 6.9) results in curves which are close to the model predictions, and which differ in both (positive and negative) directions. The too optimistic transformation via MNRU is the result of an underestimation of

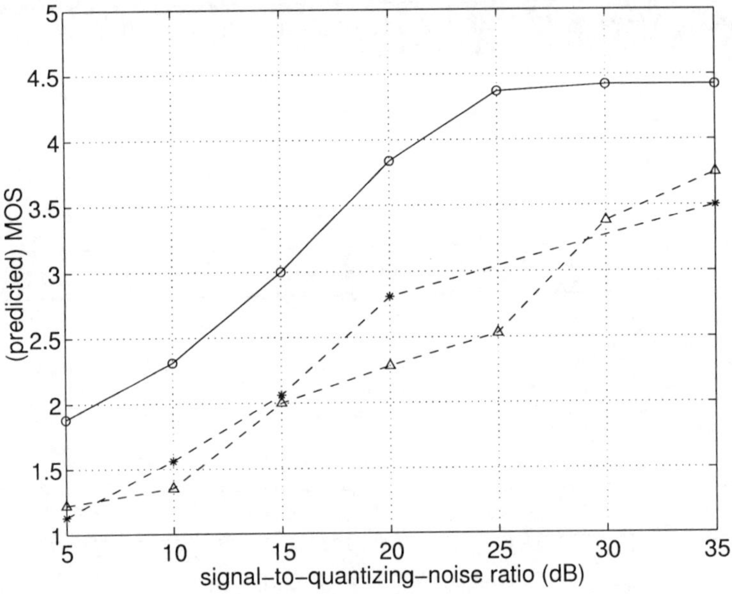

Figure 6.6. Test results and E–model predictions for MNRU distortions. Solid line: E–model prediction; slashed lines: test results (Δ: codec test, LOT; *: talker echo test, SCT).

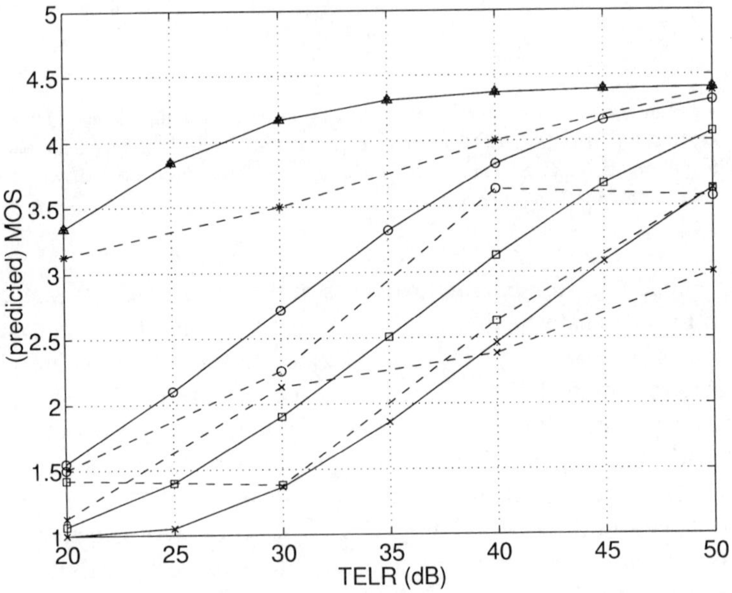

Figure 6.7. Conversation test results and E–model predictions for talker echo. Solid line: E–model prediction; slashed lines: test results (*: T=10 ms; o: T=50 ms; □: T=100 ms; x: T=200 ms).

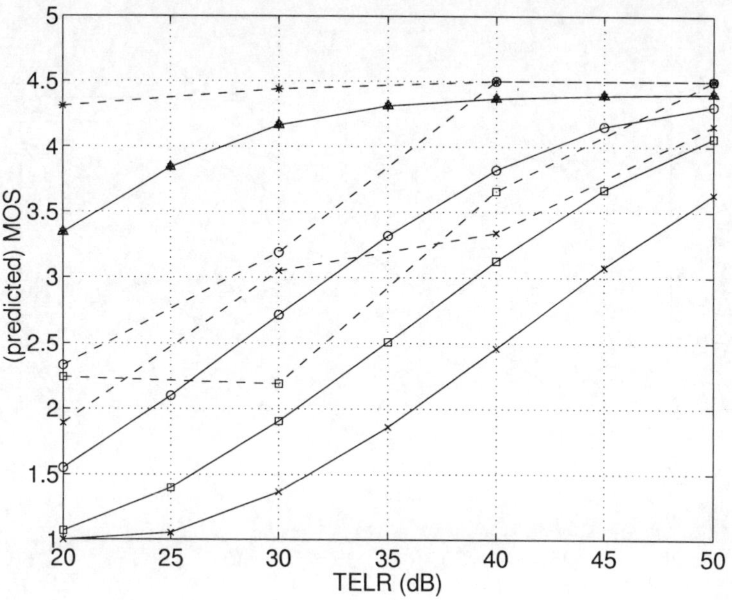

Figure 6.8. Transformation of echo test results using the "equivalent Q" method. Solid line: E–model prediction; slashed lines: transformed test results (*: T=10 ms; o: T=50 ms; □: T=100 ms; x: T=200 ms).

MNRU impairments by the E–model, see Figure 6.6 and Section 3 of Chapter 7.

For the codec listening–only test, only a transformation using the "equivalent Q" method was performed. Figure 6.10 shows that the transformation does not just lead to unrealistically positive quality estimations. The results often fall into the saturation area of the scale, so nearly all the perceived differences between codecs are destroyed.

This comparison constitutes a further argument for using similar–sounding reference conditions for comparison and transformation. The MNRU, which has been developed for degradations resulting from digital waveform equipment (PCM coding, A/D–D/A conversions), is not appropriate for performing this task e.g. in the evaluation of low–bitrate codecs, and especially not for completely different types of impairments like talker echo. Furthermore, the E–model predictions for MNRU, which form the underlying formulae for the "equivalent Q" method, should be revised in order to avoid transformation to the saturation part of the MOS scale, and this will be done in Section 3 of Chapter 7. All in all, the results of Sections 1 and 2 show that quality ratings reflecting the quality level of actual everyday situations require appropriately designed test methods comparable to the one used in the SSRCT. Context effects occur

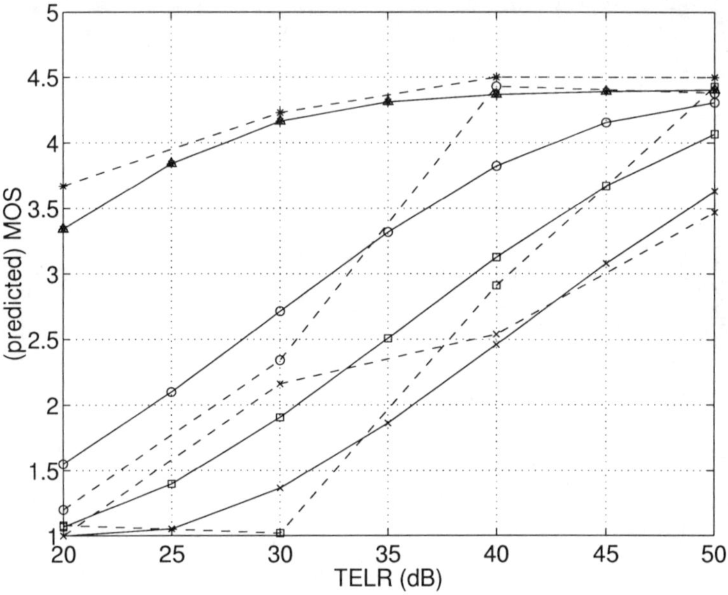

Figure 6.9. Transformation of echo test results using a linear transformation with a reference talker echo curve (T=100 ms). Solid line: E–model prediction; slashed lines: transformed test results (*: T=10 ms; o: T=50 ms; □: T=100 ms; x: T=200 ms).

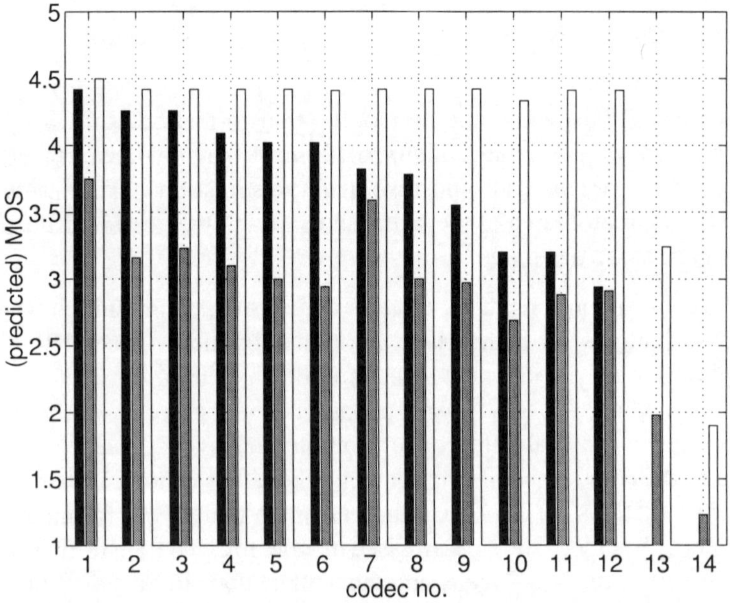

Figure 6.10. Codec test results and transformation according to the "equivalent Q" method. Black: E-model prediction; gray: original test results; white: transformed test results.

in standard listening–only as well as conversation tests, and they cannot be equalized by transformation with a reference system such as the MNRU.

3. ASSESSMENT IN LISTENING–ONLY AND CONVERSATION TESTS

Until now, little effort has been made to relate the outcome of listening–only experiments to conversation test results and vice–versa. One reason might be that it is unlikely that a generally applicable relationship between test results obtained by both methods can be found. In addition, listening–only tests are performed for class A factors (cf. Chapter 3), whereas conversation tests are used when the nature of the impairment under test does not allow the use of LOTs.

The only model which explicitly gives a relation between LOT and CT results is the British Telecom SUBMOD model. The relationship is established between the listening–effort MOS (Y_{LE}) and the conversational MOS (Y_C), by means of a function given in Section 1 of Chapter 5, Formulae 5.4 to 5.7. It is worth noticing that the relationship depends on the vocal level of the talker in the conversation situation, which is expected to be different from the one encountered in the listening–only situation. The talker is expected to raise his/her voice when the overall quality of the connection is bad, or when the overall attenuation is high, and to lower the voice when the incoming speech level is high. A further adaptation effect is due to a loud talker sidetone. The other models used for network planning purposes do not give different predictions for listening–only and conversational quality. It is not explicitly stated that the estimates are to be used in the sense of conversational quality (and in fact, many fundamental investigations rely on LOTs), but as they aim to make predictions for telephone conversations including the effects of echo and delay, it can be assumed that the predicted MOS values refer to the conversational situation. The instrumental methods presented in Section 3 of Chapter 5, on the other hand, are optimized to predict listening–only MOS from the signals used for a listening–only test, so they can only be expected to predict listening–only MOS (Y_{LQ}, Y_{LE} or DCR MOS).

In the following, results from LOTs and conversation tests using SCT scenarios are compared. All tests were carried out at IKA using the telephone line simulation system. The comparison is restricted to the conditions which were used in both tests. However, in the LOT a larger number of conditions was tested. The LOT stimulus material consisted of short text passages lasting from about 8 to 10 s, and no text was repeated for one talker. Overall, 4 talkers (2m, 2f) were used. In the SCT the number of talkers is identical to the number of test subjects, thus there is a natural variability in speech level. The conversation topic arose from the enquiries and booking situation. Besides that,

LOT and SCT were carried out over the same telephone line simulator, resulting in nearly the same physical characteristics of the connection. So listening–only and conversational tests differed mainly regarding

- the modality (listening–only or conversational),
- the topic of the speech material (purposeless vs. task–related),
- the number and choice of test conditions within one test session,
- the variability of speech levels, and
- the number of talkers.

All these factors may have an influence on the differences in test results.

In Figure 6.11 conversation and listening–only test results in terms of MOS (Y_{LQ} and Y_C) are compared for a test investigating the effect of low–bitrate codecs (left) and a different test related to the combined effects of room noise and codecs (right). As some other test conditions with room noise were included in the codec test, the overall choice of test conditions was similar regarding their perceptive effects. In both tests reference conditions resulting from MNRU distortions were included as well. In the diagrams the relation between Y_{LE} (not Y_{LQ}!) and Y_C given in the SUBMOD model is plotted as a solid line. As no speech levels were measured for the conversational situation, the difference $V_C - V_L$ was set at zero, assuming no rise in speech level due to bad transmission quality (in the auditory tests, however, the latter assumption will not be satisfied, especially for the room noise conditions).

As a general tendency the comparison shows that test subjects gave higher ratings in the conversational situation than they did in the LOT. Nearly all the data points lie above the diagonal line LOT MOS = SCT MOS. No general difference arises between MNRU, codec, low room noise and high room noise conditions, though the latter three result in ratings at different locations on the MOS scale. One main reason for the higher SCT ratings is assumed to be the test situation (modality and task) in conjunction with the choice of conditions. In a listening–only situation, the test subjects pay only little attention to what is said (if any), but mainly to the form (how it is said/transmitted). In this way the subject's attention is purely on the surface structure of the speech signal, resulting in more critical judgments. On the other hand, in a conversation subjects do not pay attention to low levels of degradation provided that good intelligibility and communicability is maintained. The large number of conditions to be tested in a LOT makes it possible to directly compare adjacent speech samples. As discussed in Section 1, testing several conditions in a close sequence does not often stimulate the test subjects to use the outer categories of the scale, so

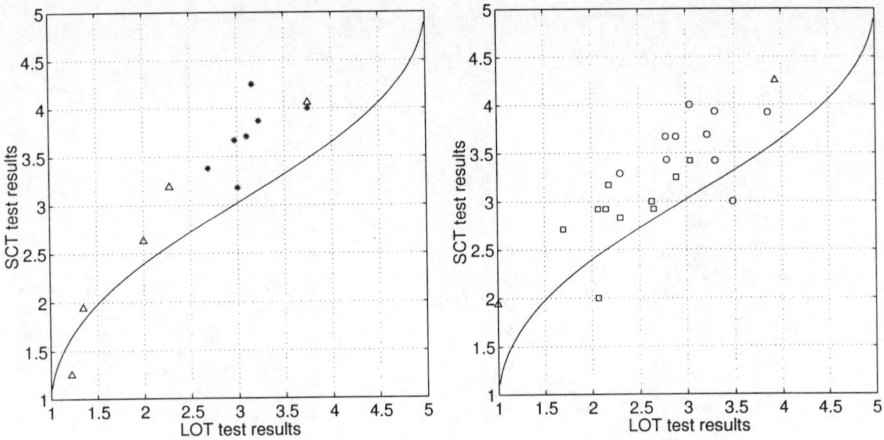

Figure 6.11. Comparison of MOS results from LOT and SCT. Left: test on codecs and MNRU; right: test on room noise + codecs and MNRU. Δ: MNRU; ∗: codec only; o: codec and room noise 50 dB(A); □: codec and room noise 65 dB(A). Solid line from SUBMOD model.

the center categories are over–represented. This effect can clearly be observed in the LOT for the upper end of the scale.

The relation between Y_{LE} and Y_C from the SUBMOD model does not help to establish a general relationship between listening–only and conversation test results. However, larger differences between test results and the SUBMOD curve are to be observed at the higher saturation edge of the MOS scale, i.e. MOS $>$ 3.5. The SUBMOD model does not take into account that these values are less commonly used in rating a larger number of connections in a sequence. Another source of deviation may be the special choice of conditions used in the test, namely the influence of room noise (for which a rise in speech level can be expected).

A somewhat different outcome allows the comparison of LOT and SCT results obtained in a test investigating the combined effects of frequency distortion and MNRU/codecs. In this test, different rating scales were used. Besides the 5–point quality scale (MOS scale) a continuous category–ratio scale (CR–10) for 'impairment' and a continuous rating scale for the 'sound impression' (cf. Figure 4.11) were used. The comparison is reproduced in Figures 6.12 and 6.13. In this test there seems to be no general difference in the quality judgments in listening–only and conversational situations. In both situations the use of the MOS scale does not seem to be restricted (though there are no test results with MOS $>$ 4.2). The same holds true for the rating on sound impression. For both scales a linear transformation line LOT MOS = SCT MOS would be appropriate. On the other hand, there seems to be a large difference in the use of the

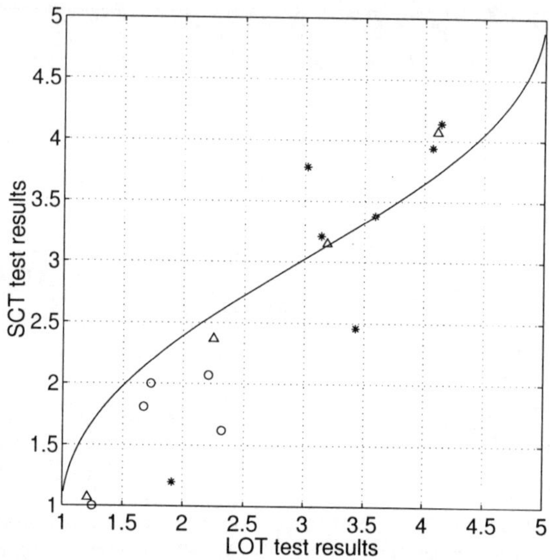

Figure 6.12. Comparison of MOS results from LOT and SCT, test on frequency distortion, codecs and MNRU. Δ: MNRU; ∗: frequency distortion only; o: codec/MNRU and frequency distortion. Solid line from SUBMOD model.

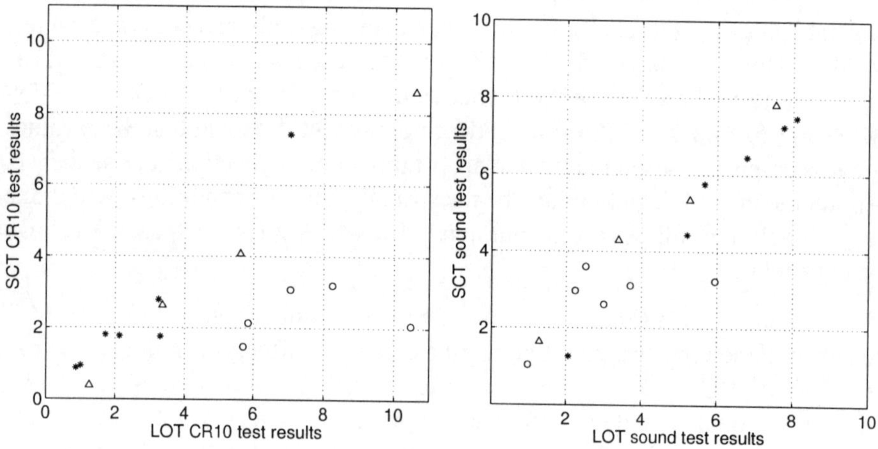

Figure 6.13. Comparison of MOS results from LOT and SCT, test on frequency distortion, codecs and MNRU. Left: CR–10 rating; right: sound rating. Δ: MNRU; ∗: frequency distortion only; o: codec/MNRU and frequency distortion.

CR–10 scale in LOT and SCT, see Figure 6.13. Here, several conditions with more severe degradations have a strong impact on the listening–only situation (high CR–10 rating), whereas the impact in terms of 'impairment' (German: "*Beeinträchtigung durch Störungen*") is lower in conversation. Once again, the focus of attention can explain the differences. It is on the content in the SCT situation and more on the form in the LOT situation, rendering the impact of degradations higher in the LOT. This difference cannot be observed on the sound scale, because the focus of the question (and the scale) relates to a very specific dimension in the speech signal.

In conclusion, no transformation law between listening–only and conversation test results can be given which would be applicable in general situations. The experimental outcome strongly depends on the quality feature to be assessed, because the focus of the assessor's attention differs with the test modality and with the scaling task. This behavior can also be seen as an argument for making use of several different scaling tasks and test modalities when more analytic information is required.

4. EXPECTATION

The term 'expectation' is a rather diffuse one, and it is not used in a unified way in telephony. Often it is used in the sense of 'advantage of access' which a user of a special system or service can experience over a corresponding standard system. An example is a mobile communication system: the user can make calls from nearly all locations (provided that an adequate coverage of the geographical area is present), and he/she is reachable for urgent calls or in cases of emergency. This advantage is not linked to characteristics of the transmission, but to the special system/service. In turn, the user may tolerate some degradations inherent in the system which would not be observable for the equivalent standard system. In this case the term 'expectation' is used as a measure of the trade–off between tolerable transmission impairments and the inherent advantage of the system.

In a more analytic picture there are three components which have an influence on the expectation defined above. These are

- the relation to the user's general experience with telephone connections (memory),
- the expectation resulting from the price of the connection or service (especially for new services there is no general equivalence "high price = high speech quality"), and
- the appropriateness of reaching specific goals, i.e. the call motivation. The appropriateness will be different for announcements or purely informative calls rather than e.g. for private calls.

It becomes obvious that handling expectation as a single cause–and–effect parameter is too simplified. As the importance of expectation will be very high for new types of services or systems (when it is not yet stabilized), it is worth describing in some detail what happens when a new product is put on the market.

The development of expectation in a new product (an innovation) can be analyzed with the help of the diffusion theory, which is generally accepted for describing consumer behavior on the introduction of an innovation. Details of this theory can e.g. be found in Wilkie (1994). Over many studies it has been found that the number of actual users of an innovation develops in an S–shaped curve, see the first diagram of Figure 6.14. The time it takes to diffuse a product depends on many factors, so no scaling can be given. Different people proceed through the adoption process at different points in time. According to the adoption time, users can be divided into 5 classes, cf. Figure 6.14, second diagram:

1. Innovators: A very small group of persons who are very quick to purchase a new product or use a service. They are very willing to accept new technologies. Innovators have been found to be people with a higher income level, higher occupational status, and they are more socially mobile than other groups. Interestingly, they are not well integrated into social groups, so they do not rely on other's opinions as to whether products suit their own purposes.

2. Early adaptors: A somewhat larger group following the innovators. They are still quick to purchase a product or use a service, but are much more integrated into their respective social groups and believe in group norms. This is an aspect which seems to be apparent e.g. for the early adaptors of mobile telephones.

3. The early majority: These people enter the market next, but they are much less willing to take risks. The group size is about one third of all adaptors.

4. The late majority: This group enters the market when 'newness' declines, so they are not really purchasing a new product or using a new service. They are less influenced by their corresponding social group behavior, and can be more easily influenced by advertisements.

5. The laggards: They enter the market when an innovation is already well accepted.

Starting off from these results of diffusion theory, a trade–off between transmission quality and user demands can be seen as the origin of an advantage or disadvantage the user of a special system or service experiences over a normal, standard system or service. In the third diagram of Figure 6.14, the potential

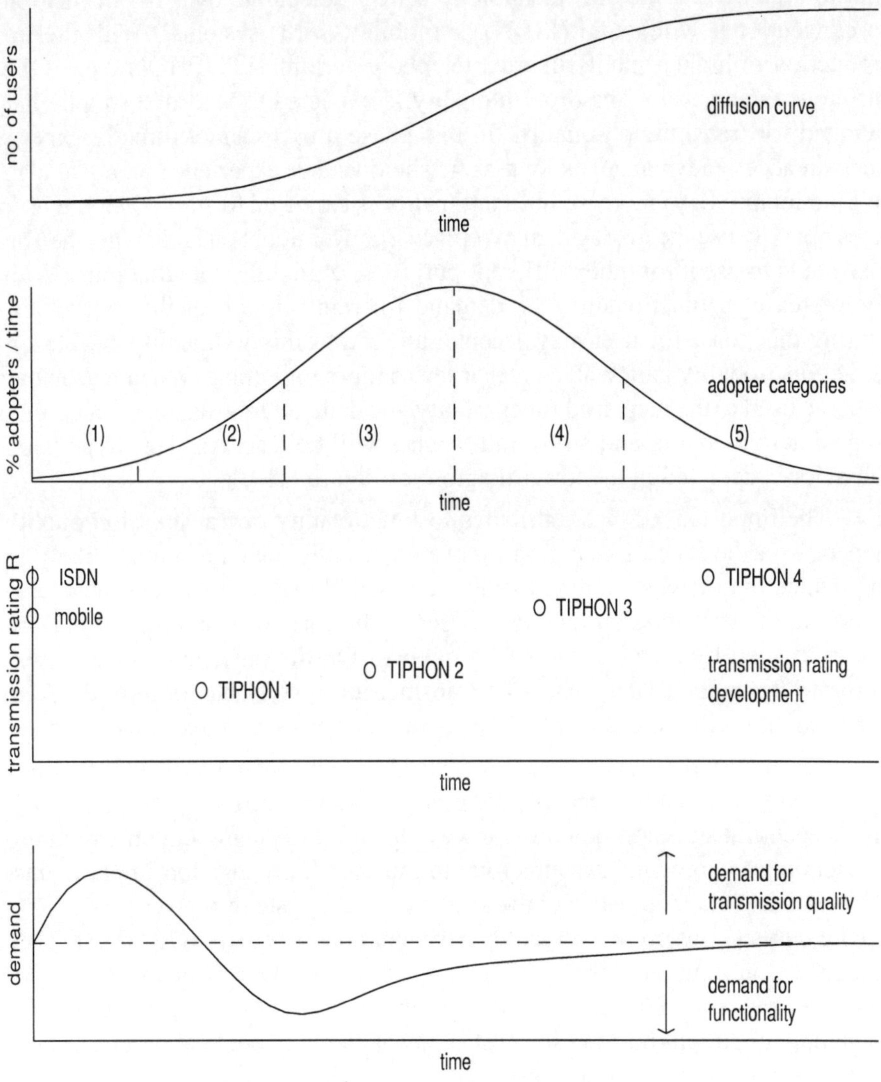

Figure 6.14. Diffusion, transmission quality and expectation for an innovation. Note: The TIPHON classes have been redefined in updates of ETSI Draft Technical Report TIPHON-05001 (1998), now based on permissible ranges for the transmission rating factor R.

transmission ratings for internet telephone terminals are given as an example, to represent a measure of the transmission quality. The values were chosen to match the minimum requirements given in ETSI Draft Technical Report TIPHON-05001 (1998) for four different Quality of Service classes (TIPHON 1 to 4). It can be seen that there is first (TIPHON 1) a clear decrease in transmission quality for the new technology called "telephone over IP" in relation to conventional wirebound (ISDN) or mobile (GSM) systems. With the introduction of higher quality internet telephone terminals (TIPHON 2 to 4) the difference decreases. The drop in quality is expected to lead first to a higher demand for transmission quality. In this phase a user cannot directly experience an access advantage, as long as he/she does not experience an advantage in functionality. An increase in functionality is expected to start when the user does not only use the new system to replace a conventional one, but when he/she also starts to use it for other, different purposes, or in different situations. With the increase in functionality, the demand for transmission quality decreases. During this phase the user may accept a lower transmission quality, because it is the functionality gain which is regarded as more important. When he/she has become used to the increased functionality, the demand for transmission quality may slowly increase, and an overall balance will be reached. This hypothetic behavior is depicted in the lower diagram of Figure 6.14.

The importance a user attributes to functionality or transmission quality depends on the type of user. An innovator or early adaptor is more likely to appreciate the increase in functionality, and will more easily accept the degradation in the transmission quality. A user of the early or late majority may be more affected by a bad transmission quality. On the other hand, these users come to the market later, when the transmission quality has improved. Also functionality will increase when a larger number of users makes this demand.

The theory developed here can be used to interpret experimental findings which have been made especially for mobile communication systems. Though the speech transmission quality is lower for mobile systems in comparison to conventional wirebound systems (due to the codec, propagation impairments, the acoustical characteristics of the terminal), these systems are very successful on the market. In a conversational experiment conducted by NORTEL (ITU–T Delayed Contribution D.009, 1997) an attempt was made to quantify the 'advantage of access' of different mobile systems in terms of an equivalent increase in signal–to–circuit–noise ratio. Test subjects had two identical wireline handsets, one simulating the traditional, wireline service and the other simulating a mobile service. The difference in the service was only introduced by describing "where" the system could be used before the test scenario started. Four different portability range categories were described: a fixed scenario (wireline service), a "home" scenario simulating a residential cordless service, a "workplace" scenario simulating an in–building wireless service, and a "national" scenario for a cellular, mobile service. Degradation in transmission quality was performed by using calibrated levels of inserted circuit noise. For both the "workplace" and the "national" scenario a detectable and consistent trade–off between voice transmission quality and (simulated) portability could be observed. From linear approximations of the functional relations MOS = $f(Nc)$ the difference was quantified in terms of a difference in line noise level. The difference was 4.5 dB for the "workplace" scenario and 8 dB for the "national" scenario. For the "home" scenario only a (statistically not significant) difference of 2 dB was calculated.

Results from codec selection tests at BT Laboratories reveal a similar observation. The performance of different codecs was assessed in LOTs using conventional fixed network telephone handsets. The overall judgment was then used to select the codec which showed the best performance. Later the codec was re–assessed in a mobile communication context which resulted in significantly higher ratings. The supposed reason was that the lower audio quality expectation in mobile communication scenarios adjusted the subjects' ratings to match their expectations (Hollier and Cosier, 1996).

In order to investigate the effect of expectation which will depend on the user group, in the context of this book a conversation experiment similar to the NORTEL one was carried out (Plücker, 1998). In each test session two subjects had 14 conversations on connections degraded by different time–invariant impairments (MNRU, codecs or talker echo). One subject used a portable, though wired handset and the other subject a standard, wireline one. Both handsets were matched in their sensitivity to give an IRS characteristic when measured according to ITU–T Rec. P.64 (1997). The SCT scenarios with the improvements for mobile systems described in Section 9.3 of Chapter 4 were used as a conversation topic. The test subject using the portable phone sat on the balcony of the institute, in other words outside the building.

The results for the different transmission conditions will not be discussed in detail. However, it can be seen from Figure 6.15 that there is no consistent difference between the mobile and the wireline handset user valid for all circuit conditions. Thus, it seems that for the specific group of test subjects and for the simulation used the expectation in terms of a trade–off between portability and transmission quality cannot be determined. This is astonishing, bearing in mind that the situation for the subject using the portable handset was much more realistically simulated than in the NORTEL experiment. A potential explanation occurs when the test results are separated for the different user groups. In this test a separation was performed on the basis of the users' experience with mobile phones. In Figure 6.16 the mean MOS results, calculated over all the test conditions, are compared for the experienced and the inexperienced users. For the experienced users there is a remarkable (though not statistically significant) 'advantage of access' of about 0.25 MOS categories. On the other side, there is a decrease in overall mean MOS of about 0.2 MOS categories for the inexperienced user group (in the NORTEL test, mean MOS differences were about 0.3 MOS categories for the "workplace" and 0.6 MOS categories for the "national" scenario). The mean ratings for the wired telephone are very similar

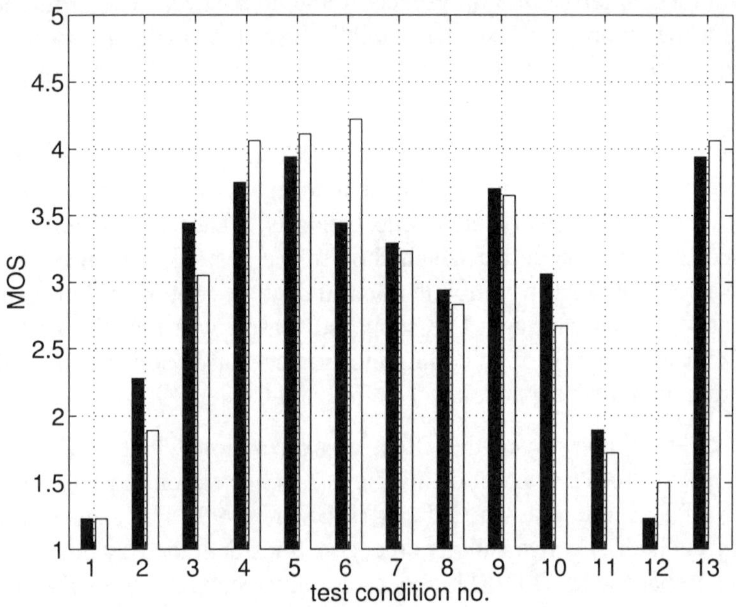

Figure 6.15. Conversation test results (MOS per test condition) for wireline vs. portable handset scenarios. Black: wireline scenario; white: portable scenario.

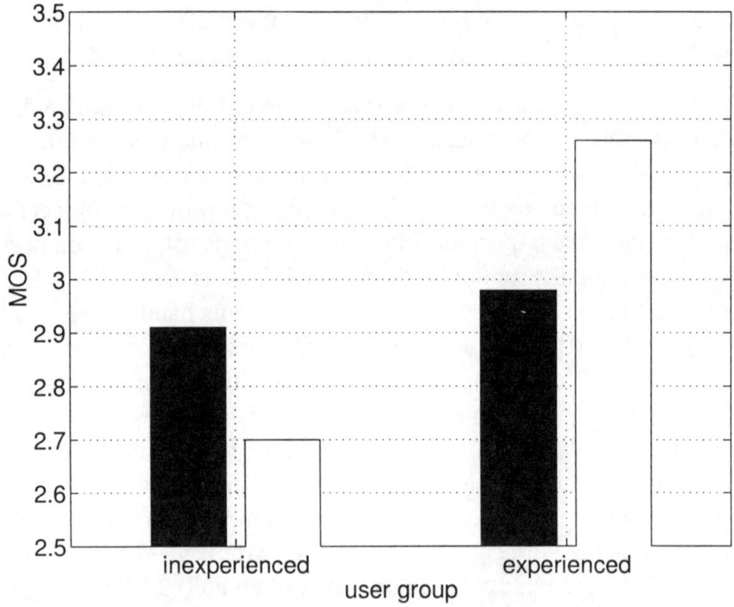

Figure 6.16. Conversation test results (mean MOS over all test conditions) for users with and without experience of mobile phones. Black: wireline scenario; white: portable scenario.

for both user groups. Obviously a potential advantage or disadvantage of access is closely related to the general attitude of the users towards the system. Among the 19 subjects of the inexperienced users' group only one was considering the purchase of a mobile phone. Many subjects of this group are expected to have a negative attitude regarding mobile phones, and several of them reported that they did not see any potential advantages of using a mobile system. On the other hand, test subjects with experience in mobile telephony reported that the transmission characteristics were in general unrealistically good, because of the absence of time–variant degradation (noise bursts, fading). It can be seen that expectation is a factor which depends to a large extent on the user group as well as on the transmission characteristics of a system.

Until now, the reported tests reflect expectation of a user confronted to a mobile handset telephone service. As a first approach in this work, a similar experiment was performed to investigate the influence of expectation on telephony from an internet terminal (Riedel, 1998). The aim was to find out whether a user who originates a telephone connection from a computer terminal has a different impression of the quality compared to a user who sets up the call from an ordinary phone. To the author's knowledge such a comparison has not yet been published in literature. A test set–up was build where both test subjects made telephone calls either from an imaginary internet terminal or from a standard phone, in each case using the same handsets. The handsets were connected to the telephone line simulator described in Section 3 of Chapter 3 which allowed transmission impairments that are typical for internet telephony to be introduced, namely combinations of overall propagation delay and low–bitrate codecs (some MNRU reference conditions were included as well). When operating the simulated 'internet terminal', the test subjects had online scenarios on the computer screen, and had to give their ratings in the screen forms. The call set–up was simulated by a commercial internet telephony software, namely Cooltalk© from Netscape. For the ordinary telephone, the original version of the SCT scenarios was used.

The results are separated according to the users' experience of using computer terminals for internet applications. None of the test subjects had already tried to use the internet for telephone connections beforehand, so it can be assumed that the subjects' experience with such a kind of service is negligible. However, it can be assumed that test subjects who are experienced with the internet may be used to longer delays. The results, separated according to the transmission characteristic conditions, are depicted in Figures 6.17 and 6.18.

It becomes obvious that test subjects with experience of using the internet (n = 12) mainly rate the connection from a computer terminal worse than the corresponding transmission over a standard telephone. The tendency is not statistically significant, but can (with two exceptions) be observed for the MNRU

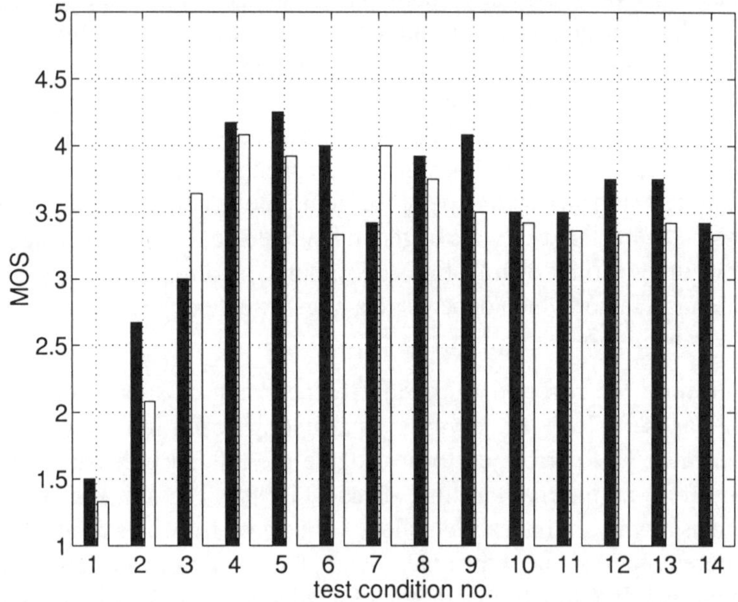

Figure 6.17. Conversation test results (MOS per test condition) for users with experience of using the internet. Black: ordinary telephone scenario; white: computer terminal scenario.

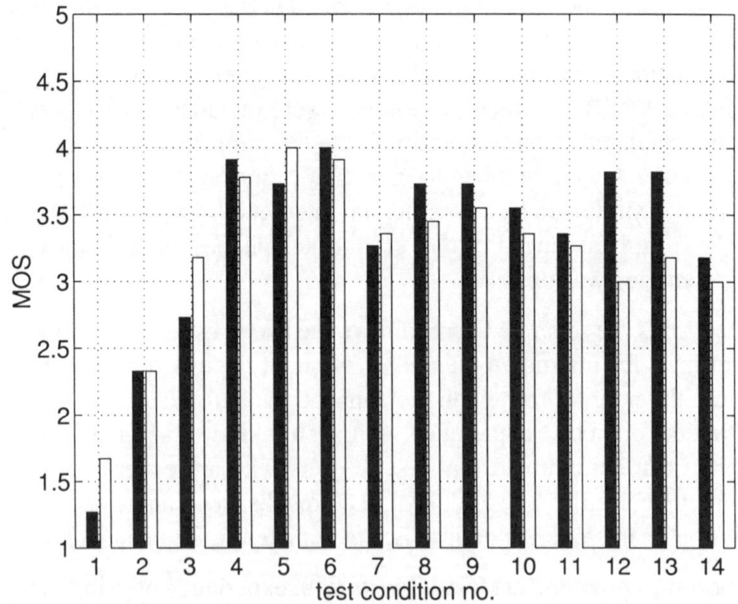

Figure 6.18. Conversation test results (MOS per test condition) for users without experience of using the internet. Black: ordinary telephone scenario; white: computer terminal scenario.

conditions (1–5) as well as for the codec and delay conditions (6–14). For the inexperienced users ($n = 11$), the tendency is similar but weaker. This is mainly due to 5 of the inexperienced subjects who have never operated a computer terminal before. This group had severe problems e.g. in operating a computer mouse, but rated the computer terminal originated calls even better for 8 of the 14 circuit conditions. For the subjects with experience of operating computer terminals, it seems that their expectation of transmission quality is more oriented towards high–quality, wideband sound (which is usual for computer sound devices). A general finding is that for both user groups the computer terminal scenario is mainly rated worse. In terms of expectation, this means that an internet terminal would now have a 'disadvantage of access', or a negative expectation factor in the E–model. This result is in accordance with the hypothesis depicted in Figure 6.14. It seems that the internet cannot incorporate an increase in functionality at the moment; it is more likely a disadvantage because the handling is more complicated for setting up a simple telephone call. On the demand curve in the figure, an internet terminal is actually located in the first part, where the demand on transmission quality is of greater importance. Because the internet cannot satisfy this demand (or at least cannot provide better quality than an ordinary telephone) it is rated worse although it has the same physical transmission characteristics.

By performing auditory tests under laboratory conditions, it was shown in this section that the term 'expectation' is closely related to the users' demands of a system or service. Therefore, it is very imprecise to make assumptions on 'advantage of access' for a general user, without specifying the particular needs and attitudes. The adopter groups defined by the diffusion theory can only give a rough classification of a specific user's demands. As expectation is to be seen in a personalized or group–related way, it is more appropriate to define 'advantage of access' trade–offs by taking into account the user's personal calling behavior. Investigations in this direction, however, have to the author's knowledge not been performed.

5. INFLUENCE OF THE COST FACTOR

Because of competition there is only very little information published on the relation between quality and price. In the past customers did not have any choice in a trade–off between price and quality. By transmitting quality information and offering pre–selection possibilities it would be possible to open this new service feature to users. One scenario is that the user can make a pre–selection of what level of quality he/she requires for a specific call, and then the call will be charged at an appropriate price. In that case it would be necessary to transmit a quality index which informs the network about the user's choice. An example of such a potential quality index is the transmission rating R, as defined for the network planning models. The factor R has the additional advantage of being additive, so that it could be calculated as a summation within the network. Proposals for such a "running quality index" have been made, but at the moment it does not seem to be of major interest to the network operators or to the users. One reason is that, in general, as far as minimum transmission quality can be guaranteed, the majority of users are not willing to pay a higher price. Exceptions from this rule may be the innovators' and early adaptors' group, see Section 4.

Pilot tests have been performed by several companies by offering a service which was paid for by an advertisement during the telephone call. The telephone call was free of charge for the caller, but the connection was interrupted every 60 to 90 seconds by an advertisement. Though no publications are available on the experiences with this kind of service, it seems that it was not acceptable to the wider public. The group it was most accepted in were young people, where the price constitutes a higher priority. For the general public the interruptions probably made the conversation unacceptable. In this case, the profit in costs did not balance the loss in conversational quality.

Due to the inherent nature of utility and acceptability investigations, in the context of this research work a different approach has to be made, namely to use auditory experiments under controlled laboratory conditions. In several conversational tests carried out in the frame of this work at the Institute of Communication Acoustics (IKA), the users were asked (apart from the quality rating) about the price they would be willing to pay for a connection of that specific quality. Two different scales were used (cf. Section 7 of Chapter 4): In a first instance, the price for a time unit for that specific connection was asked for, and a standard price of 0.12 DM was given in brackets (corresponding to the standard price of the former monopolist in Germany at that time). In a second approach a more abstract concept of "points per time unit" was used, and a standard value of 100 points was given. No specific indication about the day/night time was given. The topics of the conversations all related to the enquiries and booking scenarios. Results for two tests, compared to the respective integral quality ratings for the connections in terms of MOS, are given in Figure 6.19. The test depicted on the left side was a standard conversation test, whereas the results on the right side refer to a single stimulus conversation test. It appears that there is a strong correlation between the integral quality ratings (MOS) and the price per unit the test subjects would be willing to pay. This is not astonishing, because the price question was the last within a couple of ratings related to quality. Therefore, at least for the SCT, this behavior could be expected. On the other hand, the results of the SSRCT show a slightly curvilinear shape. It seems that there is a saturation effect at a level which corresponds to the standard price per unit a subscriber has to pay now for an ordinary domestic telephone connection. However, the results do

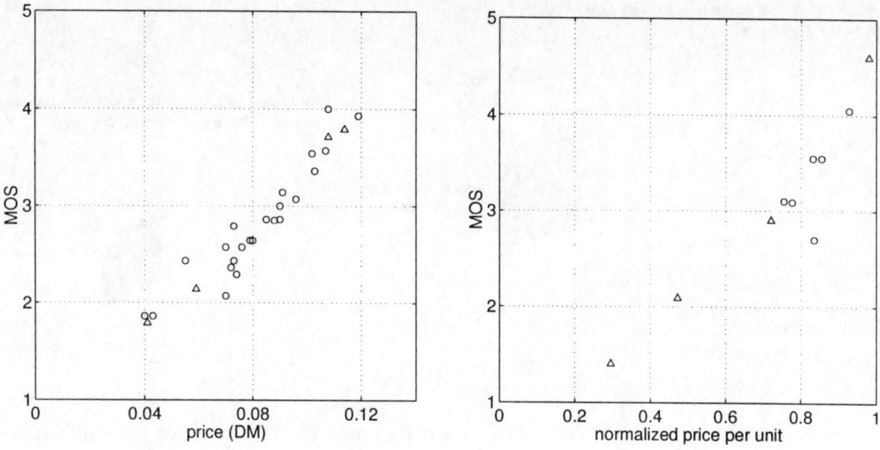

Figure 6.19. Price vs. MOS quality rating. Left: codec and talker echo test; right: SSRCT. Δ: MNRU condition; o: codec and echo condition.

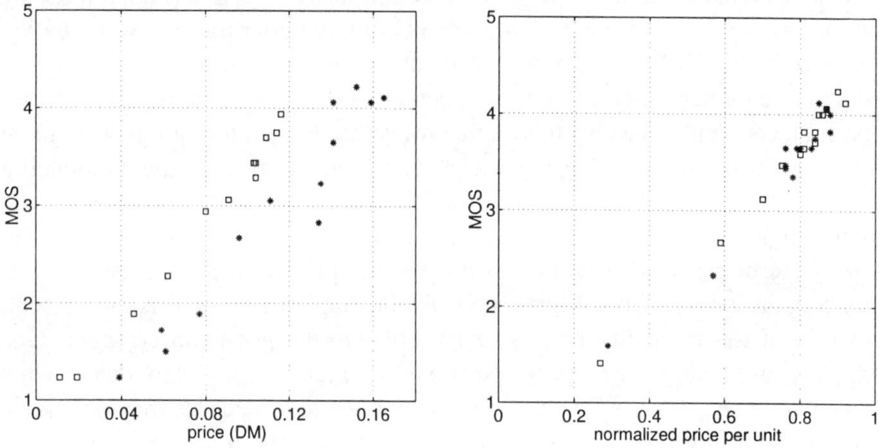

Figure 6.20. Price vs. MOS quality rating. Left: test with standard handset and portable handset; right: test with standard handset and computer terminal. □: standard handset condition; *: portable handset (left) or computer terminal (right).

not allow the establishing of a kind of threshold, which would correspond to the minimum acceptable quality. Moreover, there is no observable difference between different types of degradation. Similar results were also obtained for other tests including various impairment types.

The price a user is willing to pay for a connection will certainly depend on the type of service. For a test involving both ordinary and portable telephone

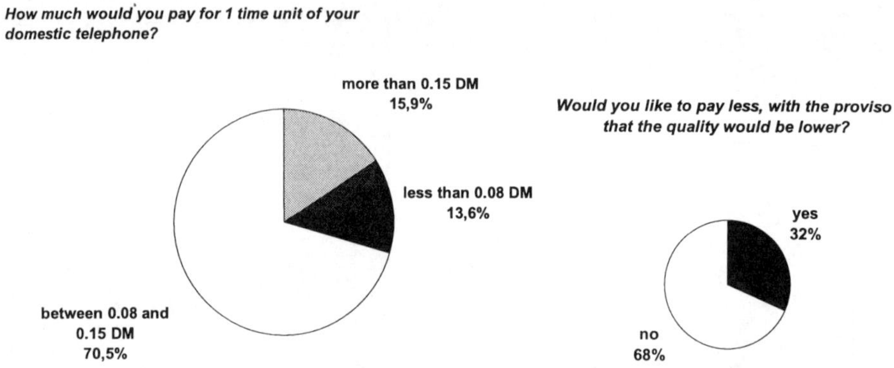

Figure 6.21. Tolerance of price in relation to quality, conversation test on frequency distortion.

handsets simulating a mobile service (see Section 4), the relation price vs. quality turns out to have a different bias, but also a linear or curvilinear shape. This is depicted in Figure 6.20. As one might have expected, test subjects are willing to pay a higher price for a correspondingly better transmission quality. The size of this difference, however, is rather large in comparison to the very slight increase in quality due to the expectation effect. For this special scenario, test subjects seem to be able to separate transmission quality ratings from price related aspects, although they are in a laboratory situation. A further influencing factor may be that the public is well aware that a mobile service will cost more than ordinary wirebound services. For the test on telephony from a simulated internet terminal (cf. Section 4) a difference in price of the two services does not become obvious, see Figure 6.20 (right diagram). There is still a high correlation to the quality ratings, but no bias in the price can be seen. One explanation would be that many test subjects either have no fixed idea of what price would be tolerable for an internet telephone (e.g. because they only have a professional internet access), or they have to pay the price for an ordinary telephone call, because they are subscribers to a private internet access service. In both cases no special price level, compared to the ordinary telephone service, would be implicitly given.

As stated in Section 7 of Chapter 4, there seems to be no published information on whether users of the standard telephone service would be willing to accept an eligible trade–off between price and transmission quality, in the sense that they could make a pre–selection of a service securing a certain transmission quality level. In a questionnaire coupled with a conversation test carried out here, subjects were asked what price they would, in general, be willing to pay for a unit of a standard domestic telephone connection, and whether they would be prepared to pay less with the proviso that the quality would be inferior. The answers have been analyzed and are depicted in Figure 6.21. It seems that the majority of the test subjects found a general price per unit of between 0.08 and 0.15 DM reasonable. About the same number of users are willing to pay more (15.9 %) or less (13.6 %). This distribution is closely centered around the general price of 0.12 DM which has been the standard price for a couple of years in Germany. Regarding the second question, it is interesting to note that more than two thirds of the test subjects ($n = 44$) were not willing to tolerate a lower quality if the price of the connection decreased. It seems that the present transmission quality is perceived as being close to the minimum threshold of what is acceptable for ordinary telephone connections (a different potential explanation, namely that the price would not be that important to the test subjects, does not seem convincing to the author). From Figure 6.3 it appears that the general transmission quality at home corresponds to the attribute "good" of a 5–point scale. From the test subjects' point of view this is also the quality which is desirable for standard connections. On the other hand, about one third of the subjects would tolerate a decrease in quality if it were linked to a lower price. It should be noted that the subjects of this test mainly originate from the student and university environment. Results may therefore be biased, especially regarding the price opinions.

6. MULTIDIMENSIONAL ASSESSMENT OF VOICE TRANSMISSION QUALITY

In literature speech communication quality and even voice transmission quality are often regarded as a one–dimensional property inherent to the speech signal or to the communication system. However, it has become obvious that this is not the case even when standardized speech samples are assessed in a listening–only situation. The reason for the simplification is that most planners of communication networks want to base decisions on one overall value of perceived quality. It is not very helpful to know that e.g. one codec produces more intelligible speech whereas speech processed by a second codec sounds more natural, if there is no fixed relation between the different dimensions and an integral quality perception.

Methods for investigating different dimensions of voice transmission quality have been discussed in Section 3.3 of Chapter 4. Several authors report assessment experiments on speech samples transmitted over communication systems or modified in other ways. Thus, the following findings relate to the multidimensional properties of one–way voice transmission quality, and to the listening–only situation. Depending on the judgment task, the kind of speech material, the assessing persons etc., some of the following dimensions were found or distinguished:

- Intelligibility or overall clarity (McGee, 1964; McDermott, 1969; Gleiss, 1970; Nakatani and Dukes, 1973; Bappert and Blauert, 1994)
- Naturalness or fidelity or speaker recognizability (McGee, 1964; Rothauser et al., 1968; Gleiss, 1970; Nakatani and Dukes, 1973; Pascal and Boyer, 1990)
- Loudness (Rothauser et al., 1968; McDermott, 1969)
- Color of sound (Bappert and Blauert, 1994)
- Distinction between background and signal distortions (Gleiss, 1970; McDermott, 1969; Nakatani and Dukes, 1973)

It is worth noting that the number of dimensions as well as their contributions to a more general judgment of integral quality, acceptability or preference is largely dependent on the test material and on the test subjects. McDermott (1969) analyzed speech samples transmitted over 22 different communication channels and thus distorted or corrupted in different ways (clipping, band limitation, noise, speech levels, vocoder speech, echo). Multidimensional scaling of preference and similarity revealed similar underlying dimensions, namely overall clarity, distinction between signal distortion and background distortion, and subjective loudness. The test subjects agreed in their rating for 'overall clarity', but differed regarding their preference for signal distortion or background distortion. Some additional variance was added by the subjective differences in preferred loudness.

Different types of speech codecs were analyzed by Bappert and Blauert (1994), using three different psychometric methods, namely magnitude estimation, polarity profile and multidimensional scaling. Magnitude estimation proved to be an adequate way of establishing rank orders regarding the overall quality, but does not give 'absolute' quality estimates nor more analytic information regarding the underlying dimensions. The polarity profile reveals two factors, an intelligibility factor loaded on the attributes "clearness", "naturalness", "nearness", "distortion" and "intelligibility", and a second factor interpreted as "colour of sound", which is loaded by the attributes "brightness", "sharpness", "contamination by noise" and "fullness". On the other side, multidimensional scaling reveals three dimensions which could not be put into relation with the factor interpretations from polarity profile. It is assumed that the dissimilarity of both results is due to the choice of particular codecs used in the test. This shows once more that there is no generally applicable method for all kinds of stimuli, but that an assessment method has to be chosen to suit both the test material and the test objective.

In order to compare different methods for assessing integral quality and more analytic characteristics of degraded speech, Panzer and Sharpley (1991) report experiments using DAM (yielding the composite acceptability measure, CA), ACR and DCR for the same set of speech samples. Degradations include MNRU, additive noise at different SNRs, and different quantization conditions. All the methods provided monotonic rank orders for each degradation type, but only between CA and listening quality MOS could a linear relationship be established. For the other assessment methods, curvilinear relationships may be assumed. The resolution power is different for each of the three methods.

McGee (1964) analyzed 15 distortions (LP, HP, BP) of one male voice by preference judgments and 30 semantic differential scales. A quality judgment scale derived from the preference judgments was related to 15 semantic differential scales and to the articulation index. The factor analysis resulted in two factors, one intelligibility factor placed through the articulation index, and one natural factor in an orthogonal direction. Both factors predicted quality preference judgments with a multiple correlation of 95 %. The perceptual spaces were further analyzed (McGee, 1965) regarding the listeners' "points of view". In the spaces corresponding to the points of view, a clustering could be observed which was interpreted to show that two distinct levels of speech quality exist for the average American listener. They correspond to the cases "low harmonics present" and "low harmonics absent". The latter result points to a problem which arises when narrow–band and wide–band telephone connections are assessed. As the quality space for the listeners might be different in each of the two cases, they should not be tested within one test requiring a unique reference for both judgments. Results from Pascal and Boyer (1990) using multidimensional analysis support this finding. In their study, similarity ratings on wideband speech samples differing in the frequency response on a simulated transmission system were subject to an INDSCAL analysis. Two factors covered 85–90 % of the variance, the first related to the presence or absence of the 100–300 Hz low frequency band and the second to the existence of the high frequency band from 5 kHz upwards. A polarity profile assessment on the same speech samples showed that two types of assessment criteria were used by the inexperienced listeners: "elementary" criteria for which intra– and inter–individual consistency is high, and speaker–dependent criteria.

7. EXISTENCE OF AN INTEGRAL "PSYCHOLOGICAL QUALITY SCALE"

Coming back to the discussion raised at the beginning of the chapter, the question remains as to whether it is — under certain circumstances — possible to establish a scale which represents a comparative measure of the trade–offs between different dimensions of speech communication quality, and their associated contribution to an overall, integral index of quality. From the experiments reported above, it will have become obvious that such a scale —

if it exists — will not be able to measure the quality perception of a specific person. It was also pointed out that there are different, sometimes independent dimensions which result in a quality event, and that there is no 'quality' which would not be linked to a specific situation. In this sense, quality can only be seen as an event, and it is generally not possible to repeat such an event, even under standardized and carefully controlled conditions.

Nevertheless, inter–individually valid quality estimates are necessary. It would be of great use for people who have to rely their decisions on quality measures or even quality estimates and predictions to express quality in terms of a single, one–dimensional parameter, which should carry the significance of an overall (mouth–to–ear) integral (carrying all potential dimensions of quality) quality index. Several requirements must be met for such an index and the related scale:

- The index should provide a comparative measure of the trade–offs between different dimensions of quality (integral quality).
- The measure should represent a mean estimate of all the potential users of a system. It cannot be expected to represent one specific user's perception of quality. This requirement can only be met if various quality measurements obtained at different locations and under different environmental conditions can be used for its construction.
- It would be desirable for the quality scale to possess some sort of additivity property. I.e., an impairment due to one factor, where there is also another factor present, should provide the same measure as when the second factor is absent. Such an additivity makes a scale particularly useful in determining overall influences of several distinct factors in combinations, and would open the quality measure up to new types of quality influencing factors.

The principle of additivity has already been discussed in Section 2.1 of Chapter 5. It was pointed out that —according to the experimental investigations described by Allnatt (1975)— the impairment ratings obtained on a ratio scale may be additive to a certain degree. Though one of the requirements was to have a scale measuring integral quality, an attempt will be made here to establish a scale in terms of 'impairment' and not integral quality. The reason for this is that impairment is expected to be a more unidimensional characteristic. The concept of impairment can also be correlated with the 'impairment factor principle' and the transmission rating scale (R–scale), which are used as the basis of transmission planning models, especially of the E–model. The transmission rating scale does possess additive characteristics by definition, and the models built on this scale make extensive use of this property.

Unfortunately, there is no method for directly assessing the influence of transmission impairments on the R–scale. Considering the drawbacks of the magnitude estimation method, category–ratio scaling seems to be an interesting alternative for forming a basis for the quality scale to be developed (cf. Section 8.3 of Chapter 4). It combines the advantages of the category scale (enabling absolute level determination) and the ratio scale (enabling the determination of relations between impairments, the main characteristic of Allnatt's scale). The easiest and most widely used version of this scale is the CR–10 scale, cf. Figure 4.12. The scale is meant to judge the magnitude of a specific perception. In our case, it is used to describe the magnitude of an 'impairment' in general terms.

The transmission rating R is defined by the E–model in Formula 5.19. Assuming the default parameter settings of a connection with good transmission characteristics (ITU–T Rec. G.107, 1998, Table 3), a simplified version can be written as

$$R = 95 - Itot \tag{6.4}$$

assuming

$$Itot = Is + Id + Ie \tag{6.5}$$

and $A = 0$.

A comparison of the CR–10 scale can now be made with the relation between $Itot$ and "speech communication quality" given in ITU–T Rec. G.113 (1996), and with the relation between R and "user's satisfaction" given in ITU–T Rec. G.175 (1997). This has been performed in Table 6.1. It can be seen that the attributes of an impairment related CR–10 scale agree very well with the E–model guidance for planning purposes. The CR–10 values and the impairment factor $Itot$ can be directly related by the linear function

$$Itot = 10 \cdot (\text{CR-10 rating}) - 5 \tag{6.6}$$

The constant shift of 5 between R–scale and the $Itot$–scale is well represented in the CR–10 rating, because the degradation due to the noise floor ($Nfor = -64$ dBmp, resulting in $Ro \approx 95$) is a noticeable one, and the CR–10 scale assumes that it will be noticed. In Borg's terms of perceived exertion, it corresponds to the "rest value". From the theoretical point of view, there is a high probability that impairment ratings obtained on a CR–10 scale are linearly (or close to linearly) related to the scale of the impairment factor $Itot$, and therefore to the transmission rating scale (R–scale).

It now has to be proven whether the CR–10 scale for rating telephone transmission impairments satisfies general requirements of scaling methodologies, namely on reliability and validity. Reliability (i.e. how well a scaling

Table 6.1. Comparison of R, $Itot$, "speech communication quality" (ITU–T Rec. G.113, 1996), "user's satisfaction" (ITU–T Rec. G.175, 1997) and the CR–10 scale.

transmission rating R	total impairment value $Itot$	"speech communication quality"	"user's satisfaction"	CR–10 scale numbers	CR–10 scale labels
100	—			0	nothing at all
95	0			0.5	extremely weak (just noticeable)
90	5	very good	very satisfied	1	very weak
85	10	good			
80	15		satisfied	2	weak
75	20	adequate			
70	25		some users dissatisfied	3	moderate
65	30	limiting case			
60	35		many users dissatisfied	4	
50	45	exceptional limiting case	exceptional limiting case	5	strong
40	55	customers likely to react strongly		6	
30	65			7	very strong
0	95			10	extremely strong
				•	maximal

method scales what it does scale) can be referred to both, within test (intra–test) and inter–test reliability. For several conversation and listening–only tests which were performed within this work, a direct comparison can be made between results obtained on a MOS scale and those on a CR–10 scale. Prior to computation, the results have to be normalized in order to obtain comparable figures. Here, the fact that the CR–10 scale measures perceptual intensities has

Table 6.2. Characteristics of the MOS scale and the CR–10 scale. Mean values obtained in different tests. $mean$: mean value of all the test conditions; min: minimum value; max: maximum value; $\varnothing$ std: mean standard deviation, calculated per condition. All ratings have been normalized prior to computation. Note: MOS and CR–10 ratings have opposite quality assignments.

test	conditions	$mean$		min		max		$\varnothing$ std	
		MOS	CR10	MOS	CR10	MOS	CR10	MOS	CR10
SCT A	MNRU, talker echo, sidetone + frequ. distortion	0.46	0.49	0.00	0.12	0.81	0.96	0.18	0.23
SCT B	MNRU, talker echo + codecs	0.51	0.50	0.06	0.23	0.76	0.96	0.20	0.23
SCT C	MNRU, codecs + delay	0.59	0.38	0.11	0.20	0.76	0.85	0.23	0.23
SSR-CT	MNRU, codecs + talker echo	0.52	0.43	0.09	0.11	0.89	0.77	0.16	0.17
LOT	MNRU, codecs + Ps	0.51	0.64	0.01	0.35	0.88	1.01	0.19	0.16

to be taken into account. The standard deviation on such a scale will increase with the scaling numbers in a close–to–logarithmic way, because the scaling numbers are chosen in this way (to satisfy the Fechner law). In addition, the overall range of the scale values has to be matched. Therefore, the following normalization rules were used:

$$MOS' = \frac{MOS - 1}{4} \qquad (6.7)$$

$$(\text{CR-10 rating})' = \frac{ln[(\text{CR-10 rating}) + 1]}{ln(11)} \qquad (6.8)$$

Using these normalized values, the following values were calculated: The mean, minimum and maximum values of all the test conditions in a test, in order to investigate the use of the scale; and the mean of the standard deviations

Table 6.3. Parallel test reliability between the MOS and the CR–10 scale. r: Pearson correlation coefficient, calculated per stimulus (SCT A – C and SSRCT) or per circuit condition (LOT). All ratings have been normalized prior to computation.

test	conditions	r
SCT A	MNRU, talker echo, sidetone + frequency distortion	-0.820
SCT B	MNRU, talker echo + codecs	-0.905
SCT C	MNRU, codecs + delay	-0.832
SSRCT	MNRU, codecs + talker echo	-0.831
LOT	MNRU, codecs + Ps	-0.970

calculated per test condition, also on a per test basis, as a measure of the inter–subject reliability. The data for five different tests are given in Table 6.2. It can be seen that both MOS and CR–10 scale are used around their center part ($mean$), and that the range of mean ratings on both scales is about the same. On the other hand, the standard deviations for measurements ($\varnothing std$) are comparable on both scales as well.

Correlations between ratings on two different scales used simultaneously in a test are one way of estimating reliability (Borg, 1998). This parallel test reliability was calculated for all five tests. In the SCTs (A–C) and in the SSRCT correlations could be calculated on a basis of each individual stimulus (i.e. conversation). For the LOT correlations were calculated on the basis of the mean ratings per circuit condition, as only one of the two scales was used by each test subject. The Pearson correlation coefficients defined by

$$r = \frac{\sum_{i=1}^{N} (x_i - \overline{x}) \cdot (y_i - \overline{y})}{(N-1) \cdot S_x \cdot S_y} \tag{6.9}$$

are listed in Table 6.3. Relatively high correlations between 0.82 and 0.97 indicate a high parallel test reliability between the MOS and the CR–10 ratings.

Regarding the inter–test reliability, for several MNRU conditions as well as one configuration using the default parameter settings in Table 3 of ITU–T Rec. G.107 (1998) the results obtained in different tests were compared. The normalized results are given in Table 6.4. For the conversation tests (SCT A to C and the SSRCT) inter–test reliability is fair, keeping in mind that the choice of test conditions as well as other test parameters are significantly different for the tests. The differences between results obtained in different tests seem to be slightly higher for the CR–10 scale and for the low–quality conditions (MNRU

Table 6.4. Inter–test reliability of the MOS and the CR–10 scale. Comparison of mean results obtained in different tests. All ratings have been normalized prior to computation. Note: MOS and CR–10 ratings have opposite quality assignments.

cond.	SCT A		SCT B		SCT C		SSRCT		LOT	
	MOS	CR10	MOS	CR10	MOS	CR10	MOS	CR10	MOS	CR10
MNRU 5	0.05	0.91			0.11	0.85			0.01	1.01
MNRU 6			0.06	0.96			0.09	0.77		
MNRU 12			0.27	0.75			0.27	0.66		
MNRU 15	0.34	0.62			0.38	0.57			0.28	0.79
MNRU 20	0.54	0.45	0.56	0.47	0.54	0.42	0.48	0.46	0.33	0.72
MNRU 35	0.77	0.12	0.73	0.25	0.76	0.20	0.89	0.11	0.88	0.35
default			0.75 0.76	0.24 0.23					0.80	0.39

5 to 12), but they are quite satisfying for the medium and high quality conditions. Results of the LOT deviate more from those obtained in conversational situations on both scales. Unfortunately no test has been repeated with the same test conditions and with the same subject group. In one test the default test condition was tested twice, once at the beginning of the test and once in different positions in the test sequence. It can be seen that almost identical results were obtained on both scales for that condition.

It is much more difficult to test the validity of the quality aspects measured on the new scale, i.e. how well the scale measures what it should measure. Some definitions of what should be measured has to be chosen. In accordance with the requirements, the quantity described by the transmission rating scale or the (linearly related) impairment scale will be chosen for this purpose. In this way, it can be determined how well both scales measure those aspects which are included in the "psychological quality scale", as used in transmission planning and in the E–model. The degree of correlation will be expressed in terms of the Pearson correlation coefficient r and the mean squared error

$$mse = \frac{\sum_{i=1}^{N}(x_i - y_i)^2}{N} \tag{6.10}$$

Table 6.5. Correlation between E-model predictions and test results on the MOS and the CR–10 scale. r: Pearson correlation coeff., mse: mean squared error.

test	conditions	MOS		CR–10	
		r	mse	r	mse
SCT A	echo + sidetone	0.8971	0.0111	0.9127	0.0358
	MNRU, echo + sidetone	0.8935	0.0107	0.8720	0.0312
SCT B	echo + codecs	0.9717	0.0077	0.9386	0.0110
	MNRU, echo + codecs	0.9523	0.0110	0.9116	0.0155
SCT C	codecs + delay $<$ 200 ms	0.9483	0.0024	0.9628	0.0042
	MNRU, codecs + delay $<$ 200 ms	0.9802	0.0088	0.9737	0.0030
SSR-CT	codecs + echo	0.8173	0.0503	0.9178	0.1153
	MNRU, codecs + echo	0.6583	0.0392	0.6257	0.0705
LOT	codecs + $Ps = 50$ dB(A)	0.9162	0.0160	0.9693	0.0160
	MNRU, codecs + $Ps = 50$ dB(A)	0.7957	0.0304	0.7479	0.0339

These values have been calculated in relation to the E–model predictions in terms of R and MOS. The results are given for the five tests cited above in Table 6.5. The test conditions were separated into those with and without the MNRU, because it turned out (see Section 3 of Chapter 7) that quantizing distortion is not well predicted by the E–model. The latter observation can be verified by the lower correlation coefficients (except SCT C) when the MNRU is included, for both MOS and CR–10 ratings. For the same reason the conditions with high ambient noise ($Ps = 65$ dB(A)) and long absolute delay were excluded. The table shows that correlations between E–model predictions and the CR–10 ratings are slightly higher in mean compared to the MOS scale, especially when the MNRU conditions are disregarded. However, variability of the correlation (mse) is higher for the R–scale. This is not astonishing, as the comparison had to be made on the basis of the untransformed values. The relations between auditory test ratings and E–model predictions for both scales are given in Appendix H.2.

The results of the validity as well as of the reliability investigation show that the CR–10 scale used to investigate telephone transmission impairments has a degree of accuracy similar to the MOS scale. In addition, it has been shown that the CR–10 ratings can be linearly transformed into quality estimates on the transmission rating scale R. The question remains as to why the CR–10 scale should be preferred to the traditional MOS scale, which has been used for a long time and under very different test conditions and in different countries. One main advantage is that the direct measurement on a scale closely related to the transmission rating factor R permits impairment factors to be directly estimated, without having to transform test results on different scales. Thus, it allows a direct estimation of new impairment factors. The CR–10 scale has a higher scale level than a traditional category rating scale, and because it is continuous test subjects are implicitly invited to make more finely graded ratings, and thus to pay more attention to what their opinion really is. On the other hand, it is highly desirable to have more data available regarding the use of the CR–10 scale in different test laboratories (language and cultural conditions), and especially to perform more investigations on test–retest reliability and intra–test reliability, both for listening–only and conversational modalities. The outcome of the latter experiments could also lead to a re–consideration of the reference and normalization problem discussed in Section 2. The anchoring of the CR–10 scale in the context of telephony is an object for further study as well. It should be noted again that the CR–10 scale is copyrighted and can only be used with prior permission of the author (Borg, 1998).

8. SUMMARY

In Chapter 2 a user–oriented schematic of factors governing the quality of a telecommunication voice service has been established, and the quality features experienced by the user have been related to elements describing the telephone line, to assessment methods investigating the quality features, and to different modeling approaches (Chapters 3 to 5). The question now arises how the interrelationship between quality features affects the outcome of assessment experiments, and of course the quality of the service in terms of usability, utility and acceptability. It has been addressed in this chapter by means of auditory experiments under laboratory conditions, because of the impossibility to carry out real user surveys and acceptability measures, but also because complete control over transmission circuits is gained. An even deeper view into the interrelationships requires the use of the latter methods on a multitude of existing telecommunication networks.

Inherent to standard laboratory tests is the difficulty to obtain quality judgments which reflect the situation in which the system will later be used. For a telephone system, the test results ideally should reflect the standard telephone situation in everyday (private or professional) life. However, different types of context effects occur in absolute category rating tests, which prevent test results from being ‘absolute’. A conversation test with single stimulus ratings (SSRCT) was carried out, and the results obtained were compared to SCT test results. The SSRCT overcomes the limitations of standard tests, however at high time expenses. It also reveals that the most popular MOS scale is used in an non–symmetrical way: the standard telephone connection results in an image in the upper half of the scale. Context effects are recommended to be neutralized by transformation of the test results with those obtained for a reference degradation. Several reference degradation units were presented, which allow a continuous manipulation of the amount of degradation introduced into the connection. It was discussed and verified by means of several auditory tests that the perceptive nature of the degradation decides on the success such a normalization may provide. No general transformation methodology will render test results obtained in one specific laboratory test independent of the test situation, context effects, etc. On the contrary, the uncritical use of transformation rules can destroy the informative outcome of an experiment.

Telephone connections will most often be used in a conversation mode. Nevertheless, LOTs are popular because they permit the testing of more different circuit conditions within one test session. The differences in quality judgments between LOTs and CTs do not only reflect the greater variety of voices and speech levels inherent to conversation tests, but also the judging person's attention. Whereas in a listening–only situation the listener's attention is more directed towards the form of the transmitted speech signal, in a meaningful conversation the function of speech is stressed. As a result, in many conversation tests subjects give higher ratings in average than in LOTs. However, no unique relationship between LOT and CT MOS results can be established which would have validity beyond the specific test it has been derived from. The analysis of auditory test results performed in this work further shows that the relationship depends on the quality dimension to be assessed, as well as on the circuit and environmental noise condition. It supports the necessity of using conversation tests (preferably SCTs) in order to achieve veridical conversation efficiency measures.

Quality is the result of a comparison process between a percept and the expectation and desires of the perceiving person. Thus, user expectation will have a major influence on perceived quality. In the context of telecommunication services, the term expectation is more often used in the sense of 'advantage of access' which a user of a special system or service experiences over a corresponding standard telephone system. Two examples were addressed by means of own auditory tests: a mobile system and telephone conversations carried out from a computer terminal. Differences in the overall quality judgment can be observed between those systems and standard ones. However, the amount of the corresponding advantage or disadvantage of access depends on the user group, and the specific desires and motivations of this group. The fact has been illustrated with the help of diffusion theory used in marketing. For mobile systems, a slight advantage of access could be observed for experienced users. Telephony from an internet terminal, on the contrary, may show a 'disadvantage of access' as long as no additional functionality is apparent to the user.

In terms of the QoS schematic defined before, the costs of a telecommunication service carry an influence on its utility, and consequently on the acceptability. These quality features are difficult to investigate by means of tests carried out under laboratory conditions. Nevertheless, some results from own auditory tests were presented, which allow to relate integral quality judgments (MOS) to the price a user would be willing to pay for a connection of specified quality. Linear or slightly curvilinear relationships were obtained. For different systems (e.g. conventional vs. mobile systems) the relationship is biased as a result of user expectation towards and experience with a system.

Even one–way voice transmission quality is multidimensional in nature. Several investigations cited from literature were analyzed, showing that intelligibility or overall clarity, naturalness or fidelity or speaker recognizability, loudness, color of sound, and the distinction between background and signal distortions are among the relevant dimensions. For practical applications, however, it is desirable to establish a quality scale which represents the trade–offs between different dimensions of speech communication quality, in terms of an integral mouth–to–ear quality index. A proposal was made for the measurement of such a quality index, related mainly to the transmission rating scale which is at the basis of the BcTR and the E–model. Category–ratio scaling is used to establish the relation. In turn, the new method allows direct measurements of impairment factors to be obtained. It was tested in several experiments (SCTs, LOTs and a SSRCT), involving a great variety of different circuit and ambient noise conditions. Parallel test reliability between MOS and category–ratio ratings is high, and inter–test reliability is comparable for the MOS and for the CR–10 scale. The validity of the scale was checked in terms of an agreement with the E–model definition of impairment factors. Higher correlations to E–model predictions were obtained on the impairment scale than on the MOS scale. The proposed methodology thus presents the first approach to perform direct auditory measurements on the internal scale which is on the basis of modern transmission planning.

Chapter 7

QUALITY OF PREDICTION MODELS

From the last chapter it has become obvious that a number of different factors have an influence on the quality of a service, and that it is not possible to make accurate and valid quality measurements without taking these factors into account. Therefore, auditory testing is required to measure quality. On the other hand, in the planning and set–up of telecommunication networks it is very often not possible to perform auditory tests before the network has been established. The requirement of implementing high–quality systems and services makes it necessary to have adequate planning tools on hand which can reasonably estimate integral quality as it would be perceived by a general user of the ensuing service. Different approaches for making such quality predictions were discussed in Chapter 5.

It is now worth investigating how well such prediction models can estimate speech communication quality, or some of the entities related to the quality of a service, i.e. to estimate the quality of prediction models. Especially the two models which have been discussed in more detail, the E–model and the SUBMOD model, have a large number of input parameters. Thus, it is not possible to analyze all the potential combinations of parameters. Within the last years, the E–model has been thoroughly tested (Möller and Jekosch, 1998; Bodden and Jekosch, 1996), and a number of test results have been published within the ITU–T and elsewhere (ITU–T Contribution COM 12–37, 1997). These results will not be reproduced here. However, the impact of an exemplary choice of transmission parameters will be discussed in more detail in the following sections. The choice was made in order to show at which points the quality of prediction models is not satisfactory and can still be improved. Especially those parameters which are of great importance for modern telecommunication networks were investigated. The analysis is done in principle by comparing model predictions to auditory test data obtained during the work at IKA, and using the telephone line simulation described in Section 3 of Chapter 3. Comparison will be made mostly to the E–model (which is the only standardized network planning model at the moment), and to the SUBMOD model, where appropriate. In Section 4 instrumental models making quality predictions based on measured signals (TOSQA and PSQM) will be used as well.

1. PREDICTION FOR SINGLE PERCEPTIVE TYPES OF IMPAIRMENT

In this section the quality predictions on speech communication systems which are degraded by one specific perceptive characteristic of impairment will be discussed, with the exception of impairments related to speech codecs. The latter will be discussed in Section 3. When performing own experiments, all the remaining parameters of the simulation can be set to their default values, e.g. those given in ITU–T Rec. G.107 (1998), Table 3, and the influence of only one single parameter (or a pair of parameters, e.g. talker echo delay and attenuation) is assessed. However, in experiments performed at other locations —and for other purposes— often not all the parameter settings necessary as an input to transmission planning models have been measured. In that case, default assumptions have to be made for the remaining parameters. The case of different degradations occurring simultaneously will be discussed in Section 2.

In ITU–T Contribution COM 12–37 (1997) the author performs an analysis of the E–model sources. It turns out that for the traditional degradations overall loss and circuit noise the E–model is based on the Bellcore model (BcTR), however with slight modifications to the optimum loudness. The latter is a function of the circuit noise, a fact which was not modeled by BcTR. No new test results on these traditional transmission parameters have been reported since then, and it is assumed that the E–model gives reasonable quality estimations. Slightly contradictory results were obtained for the prediction of the effect of talker sidetone. Whereas a good agreement between E–model predictions and a Standard Telecom Lab. test was found (ITU–T Contribution COM 12–37, 1997), a comparison to a test performed by AT&T (ITU–T Contribution COM 12–158, 1983) highlighted major deviations. These differences, however, may be due to an incorrect adjustment of the circuit noise. Whereas in the AT&T test two circuit noise parameter settings of $Nc = -75$ dBm0p and $Nc = -60$ dBm0p are given, the E–model default noise floor level of $Nfor = -64$ dBmp results in nearly no differences between the predicted quality curves. It can be assumed that the default value for $Nfor$ given in ITU–T Rec. G.107 (1998) is far too pessimistic, resulting in low sensibility of the model to low impairment levels. The model could be improved by assuming a lower value (e.g. $Nfor = -70$ dBmp); however, in this case the saturation effects of the (predicted) MOS scale will become more important, cf. the discussion in Section 3.

In the analysis of Bodden and Jekosch (1996) it turned out that E–model predictions for room noise at the send side and the receive side differ significantly from what can be observed in auditory tests. As already stated in Section 1.7 of Chapter 3, the content of the information in the ambient noise signal may have an influence on the overall quality judgment of the listener. A common finding in several tests regarding the effects of room noise is that for low levels of room noise test results agree quite well with E–model predictions, whereas for higher noise levels (Ps or $Pr > 55 \ldots$ 65 dB(A)) the E–model is too pessimistic (ITU–T Contribution COM 12–37, 1997). For the room noise at the receive side, it can be assumed that this may be partly due to the test subjects' quality assessment task itself. The investigations of McDermott (1969) showed that test subjects are able to distinguish between impairments resulting from a bad transmission and those that have the origin in their own environment.

Regarding the effect of room noise at the send side, the Lombard effect seems to play a major role. The E–model formulas covering the effect of room noise at the send and receive sides are based on the BcTR model; however, the effect of room noise at the send side, Ps, is not directly covered by the Bellcore model. It can be calculated in a similar way to the one for Pr, replacing the effective room noise Pre (Formula G.4) with $Ps - OLR - Ds + 21$. Though the exact origin of this relation is not clear, it shows that there is a strong link between the effects at the send and receive sides. Apparently the Lombard effect (which only occurs at the talker's side) is not modeled on the actual version of the E–model. This leads to major deviations from what can be observed in reality and in conversation tests. In a contribution by ERICSSON (ITU–T Contribution COM 12–79, 1996) it is noted that the Equation G.3, which is the respective formula to calculate the equivalent circuit noise caused by the room noise Ps, contains both a linear and a quadratic term. It further states that "the quadratic term represents a correction for the subjective effects of high noise levels...".

For modeling the Lombard effect within the E–model it is assumed that a talker speaking in the noisy environment is able to compensate for half of these "subjective effects of high noise levels", in decibel units. Hence, the supplementary amount of equivalent circuit noise Nos caused by the quadratic term should be divided by two. The corresponding E–model Formula G.3 should be changed to

$$Nos = Ps - SLR - Ds - 100 + 0.004\,(Ps - OLR - Ds - 14)^2 \quad (7.1)$$

By making this simple modification the E–model can be improved to incorporate the subjectively important Lombard effect. The improvement in prediction quality can be seen in Figures 7.1 and 7.2. The test results taken from the ITU–T documentation (ITU–T Contribution COM 12–79, 1996) are expressed

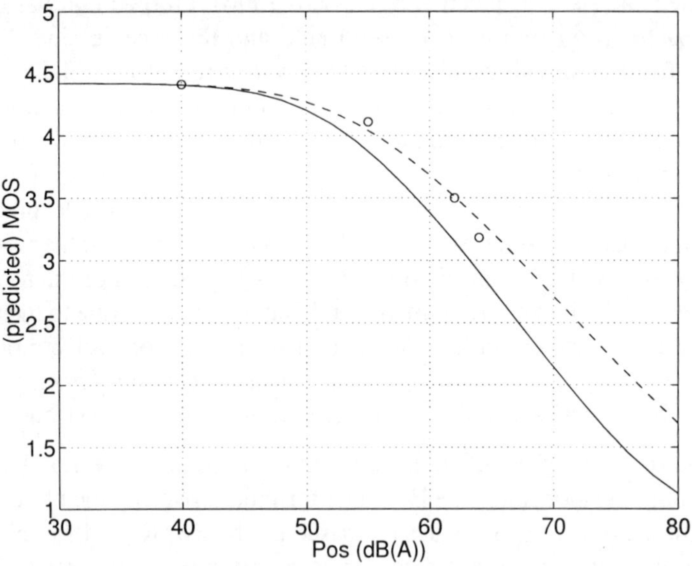

Figure 7.1. Comparison of the actual E–model (solid line), the modified version including Formula 7.1 (dashed line) and test results reported in the ITU–T documentation (o). Test parameters: $SLR = 7$ dB, $RLR = 3$ dB, $TELR = 25$ dB, $WEPL = 20$ dB; all other parameters according to ITU–T Rec. G.107 (1998), Table 3.

in values of Nos and have to be translated into room noise levels in dB(A) as used in the E–model. From Figure 7.1 it appears that the modified version of the model results in a flatter curve for the effects of Ps. This behavior correlates with the test results.

In 1995 a listening-only test was carried out at IKA. Different types of room noise were applied at the send and receive sides, and a continuous 5+2 point scale was used. As the speech files were recorded with the speaker actually sitting in a noisy environment, the test results reflect the Lombard effect as well. For the purpose of this comparison, the experimental results with $Pr = 40$ dB(A) and different levels of noise at the send side Ps are reproduced in Figure 7.2. For clarity reasons only the white noise and the 'station' noise types are given. However, similar results were also obtained using very different types of noise. Once again the modified version of the E–model, which reflects the Lombard effect, better predicts the softer slope of the overall quality curve due to Ps. The deviation for low levels of noise may have two different sources: the overall noise floor $Nfor = -64$ dBmp which, in conjunction with a high loss connection ($OLR = 20$ dB) results in a more severe degradation, and a scaling problem. Further comparisons to test results obtained at CNET as well as an own test on combined effects of room noise and codecs support the view that the modification of Equation 7.1 improves

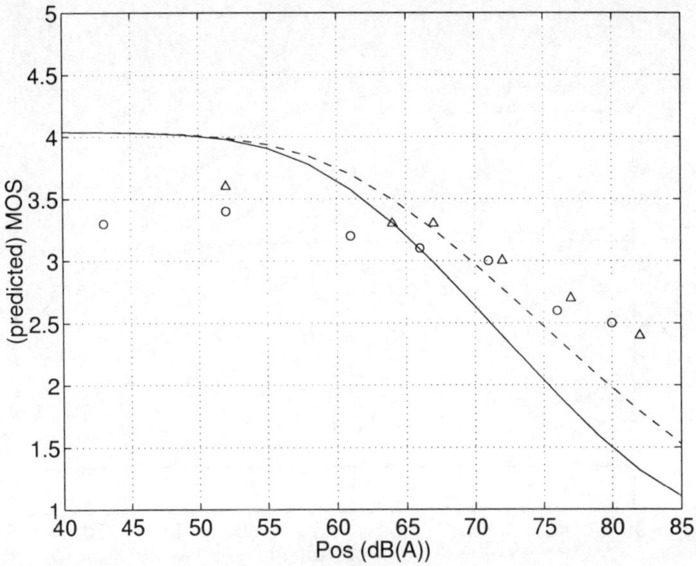

Figure 7.2. Comparison of the actual E–model (solid line), the modified version including Formula 7.1 (dashed line) and own SCT results (o: white noise, Δ: station noise). Test parameters: Pr = 40 dB(A), $SLR = RLR$ = 10 dB, $STMR$ = 12 dB, $LSTR$ = 15 dB, $qdu = 0$, $WEPL$ = 80 dB; all other parameters according to ITU–T Rec. G.107 (1998), Table 3.

the E–model predictions for room noise at the send side (ITU–T Contribution COM 12–70, 1998).

The effect of room noise at the send and receive sides has been the object of a listening–only test carried out at IKA. The test results for hoth type noise (modeling "office" type noise according to Hoth (1941)) are depicted in Figure 7.3, but very similar results were obtained for "station" type noise and speech babble. It proves that the E–model is too optimistic for low levels of room noise and too pessimistic for high levels. The shift for the high levels of Ps is caused mainly by the Lombard effect, and the modification of the model according to Formula 7.1 reduces the difference, see Figure 7.4. For low levels of Ps the differences will be due to the MOS scale itself. In that listening–only test, the upper edge of the scale was rarely used, and even the best test condition (with $Nfor = -70$ dBmp) was rated only with MOS = 4.3. As already stated above, the high noise floor is a further reason for the low ratings for standard connections.

The SUBMOD model also takes into account effects of room noise. A comparison with the test results obtained in this work is depicted in Appendix H.3. It turns out that SUBMOD predictions are almost identical for different levels of Ps. This contradicts the test results and shows that the effect of room noise

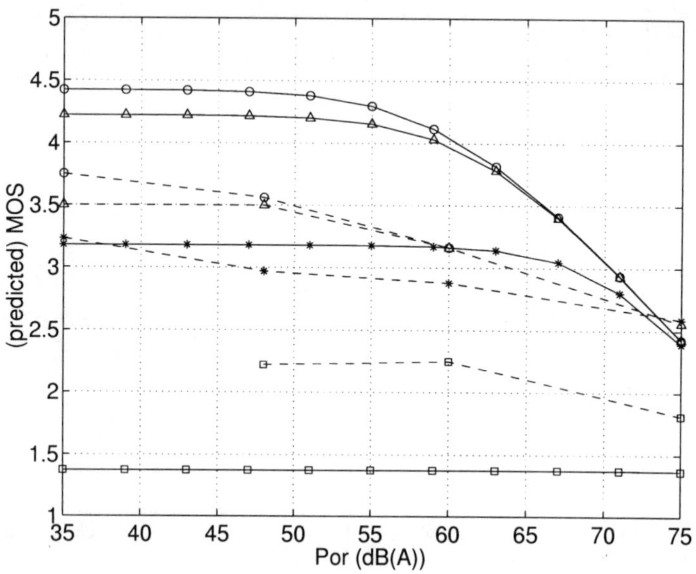

Figure 7.3. Comparison of E–model predictions (solid line) and LOT results (dashed line) for room noise at the send and receive sides (hoth noise). o: $Ps = 35$ dB(A); Δ: $Ps = 48$ dB(A); *: $Ps = 60$ dB(A); $\square$: $Ps = 75$ dB(A). Test parameters: $D = 1$, $LSTR = 16$ dB, $TELR = 50$ dB, $WEPL = 80$ dB; all other parameters according to ITU–T Rec. G.107 (1998), Table 3.

at the send side is not covered by the model. The degradation due to increasing levels of Pr, on the other hand, is predicted fairly well. Once again the predicted overall quality level seems to be too optimistic.

Talker echo due to single reflections at one subscriber's hybrid is a major degradation occurring in modern, mixed analog–digital networks. A conversation test for a large range of $TELR$ and T was carried out here, using the SCT scenarios described in Section 9 of Chapter 4. The results are given in Figure 7.5. There is a good overall agreement between E–model predictions and conversation test results, with the exception of two dropouts, namely $T = 100$ ms/$TELR = 30$ dB and $T = 50$ ms/$TELR = 50$ dB. The low MOS for the latter case might also be due to the bad resolution of the scale at high quality levels. The curve for low–delay talker echo ($T = 10$ ms) seems to be unrealistically curvilinear. However, at this level the characteristics of the MOS scale does not permit a decision on the quality of the model. Further results from NTT and CNET have been compared to E–model predictions for talker echo (ITU–T Contribution COM 12–37, 1997). Overall these comparisons support the view that the E–model does predict the effects of talker echo satisfactorily.

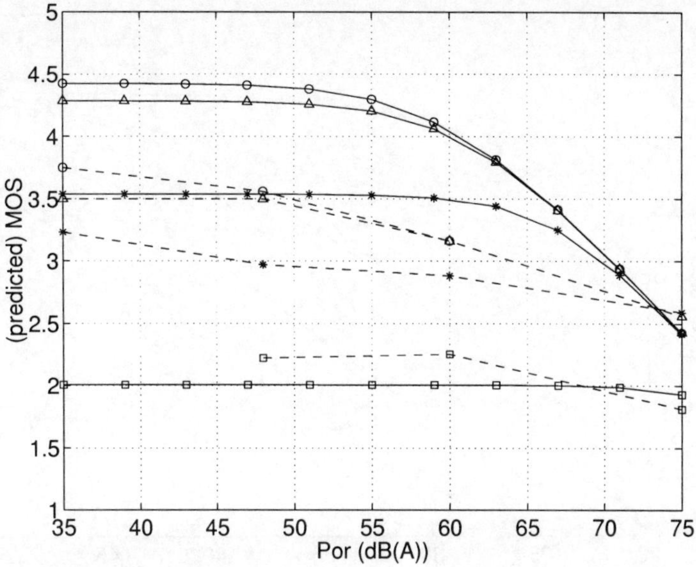

Figure 7.4. Comparison of the modified E–model using Formula 7.1 (solid line) and LOT results (dashed line) for room noise at the send and receive sides (hoth noise). o: Ps = 35 dB(A); Δ: Ps = 48 dB(A); *: Ps = 60 dB(A); □: Ps = 75 dB(A). Test parameters: $D = 1$, $LSTR = 16$ dB, $TELR = 50$ dB, $WEPL = 80$ dB; all other parameters according to ITU–T Rec. G.107 (1998), Table 3.

A comparison of the same test results to SUBMOD model predictions is given in Appendix H.3, Figure H.7. The correlation is very high for low echo delay (T = 10 ms), but becomes lower with increasing delay T. It seems that the E–model does predict the effect of long–delay talker echo more accurately. Unfortunately the exact coverage of talker echo in the SUBMOD model has not been published yet. From the description of the Call Clarity Index (cf. Section 5 and ITU–T Delayed Contribution D.088 (1998)), which is based on the same model principle, it seems that a single multiplier modifies the final value of Y_C. For the calculation in Appendix H.3 a software implementation of the complete model, kindly provided by British Telecom, was used.

Listener echo is far less of a problem in modern networks. Results from listening–only tests (Bodden and Jekosch, 1996) support the view that the E–model predicts the effect due to multiple reflections in a realistic way. The comparison is made in ITU–T Contribution COM 12–37 (1997) and will not be discussed here in detail. For that test, relatively high overall attenuation values were used, and a high quality handset may have made the degradations more noticeable. This fact can explain some larger deviations in long delay values Tr.

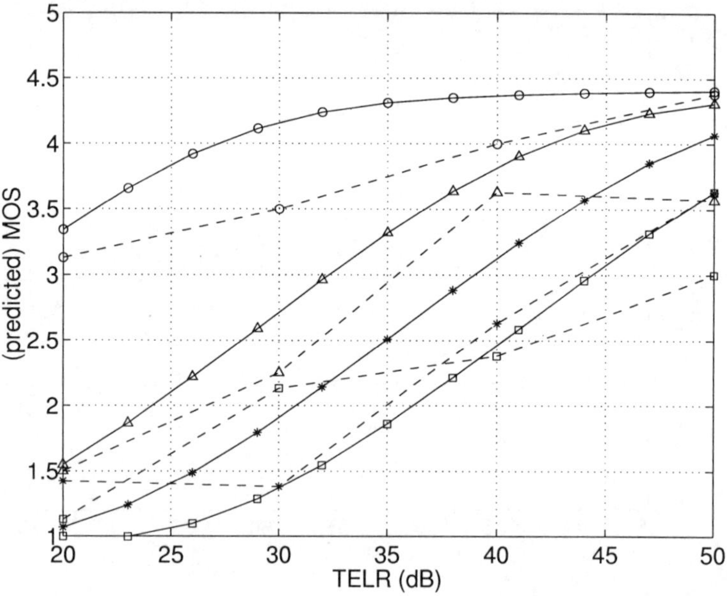

Figure 7.5. Comparison of E–model predictions (solid line) and SCT results (dashed line) for talker echo. o: $T = 10$ ms; Δ: $T = 50$ ms; *: $T = 100$ ms; $\Box$: $T = 200$ ms. Test parameters: $D = 1$, $LSTR = 16$ dB, $WEPL = 80$ dB; all other parameters according to ITU–T Rec. G.107 (1998), Table 3.

One effect which can only be observed in realistic conversation situations is the one due to absolute delay. Several test results in this regard are given in ITU–T Rec. G.114 (1996). The E–model curves for this parameter were chosen in order to fit this information, with an emphasis on highly interactive tasks and delay times up to 800 ms. Results from two tests performed by Bellcore and NTT are compared to the model in Figure 7.6. The results have been linearly transformed in order to match the E–model prediction in the absence of delay. Though it seems that the model provides a good compromise for the results, care is necessary when interpreting the NTT test results (Kitawaki and Itoh, 1991). In that test only four expert subjects made the assessment, and these subjects were trained to recognize delay in a conversation. Because of this, test results may have overestimated the annoyance of absolute delay. This assumption was verified in a later test at IKA. It is also supported by the experiments carried out by Karis (1991), where only a very slight decrease in overall quality ratings (0.1 MOS categories for $Ta = 600$ ms) was reported. The E–model predictions for absolute delay should therefore be seen in the light of the special conversation type used for its establishment. For normal, free conversations the degradation may be much less noticeable, especially when the conversation partners do not know each other very well. In that case, the effect

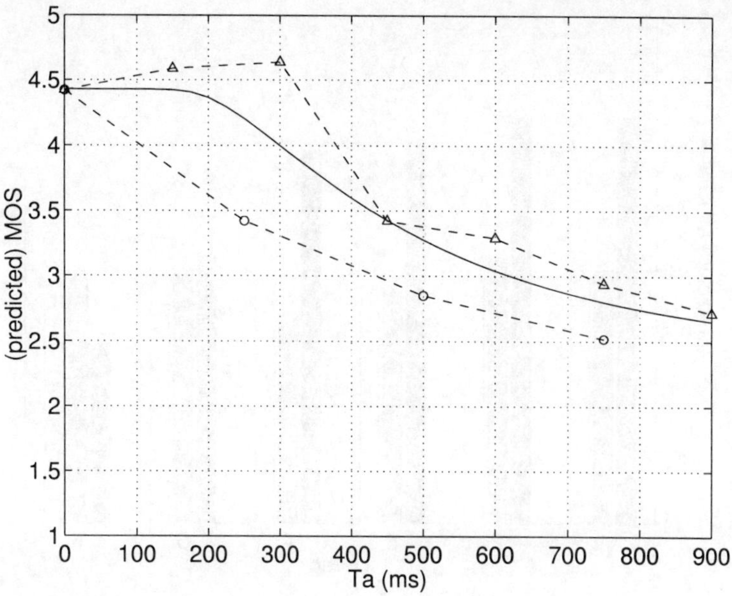

Figure 7.6. Comparison of E–model predictions (solid line) and conversation test results (dashed line) for absolute delay. o: Bellcore test; Δ: NTT test. For all parameters the values according to ITU–T Rec. G.107 (1998), Table 3, were used.

will be attributed to the opposite partner in terms of a "long reaction time", as some test subjects reported. In interviews carried out after the testing of delay–affected connections subjects indicated that polite people do not interrupt the conversation partner (ITU–T Contribution COM 12–94, 1991). The effect on the integral quality judgment, however, turned out to be very weak.

2. PREDICTION FOR COMBINATIONS OF DIFFERENT TYPES OF IMPAIRMENT

In the last section, degradations of a single perceptive characteristic have been regarded mainly in isolation (although default values for the other parameters had to be chosen). Unfortunately only a few test data are available on the combined effects of different types of impairments, e.g. room noise and echoes, codecs and room noise, etc. In the context of the E–model it is assumed that different types of impairments are mainly independent of each other, and that the corresponding impairment factors are additive on the scale of the transmission rating factor R. It has been shown (Bodden and Jekosch, 1996), however, that there are internal interdependencies resulting from the calculation of the corresponding impairment factors. As an example, the impairment factors for talker echo ($Idte$) and for environmental noise (via the basic signal-to-noise

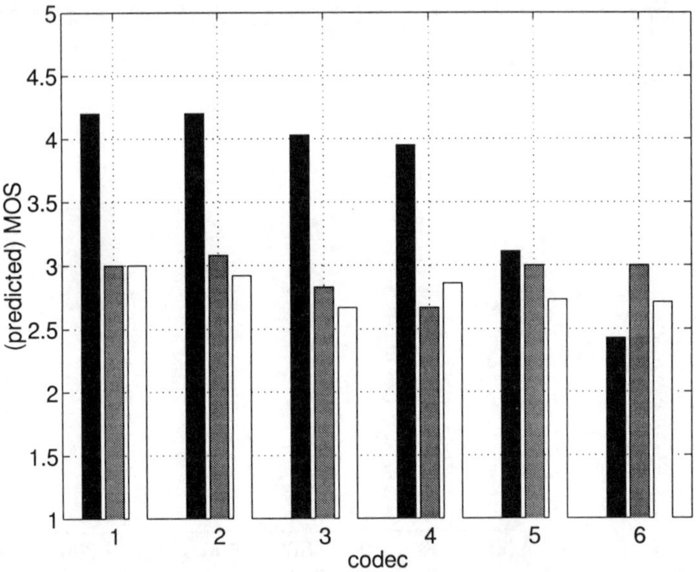

Figure 7.7. Comparison of E–model predictions (black) and SCT results regarding different codecs and $Pr = 50$ dB(A). Gray: hoth noise; white: speech babble. Test parameters: $D = 1$, $LSTR = 16$ dB, $TELR = 50$ dB, $WEPL = 80$ dB; all other parameters according to ITU–T Rec. G.107 (1998), Table 3.

ratio Ro) are related by means of the quantities No and RLR. On the other hand, there is no interdependence between the equipment impairment factors for low–bitrate codecs (Ie) and any other impairment factor. That implies that the effect of a certain speech codec on quality perception would be independent e.g. from the room noise level.

Two conversation tests have been carried out here, in which, amongst other things, different low–bitrate codecs (test A) and different talker echo attenuation and delay (test B) were investigated under noisy room conditions. Two types of noise were used, at levels of 50 dB(A) and 65 dB(A): the hoth type noise, and a speech babble. Both noise types are homogeneous regarding their temporal structure. For test A, Ps at the send side was controlled to be below 35 dB(A). For test B, symmetrical room noise conditions ($Pr = Ps$) were used. 12 (test A) and 14 (test B) subjects participated in the test. For different reasons, all the conditions including the VSELP (IS–54) codec were rated by 7 test subjects only. However, no influence on the standard deviation of the MOS values was found.

The results of test A are reproduced in Figures 7.7 and 7.8 for $Pr = 50$ dB(A) and $Pr = 65$ dB(A) respectively. It can be seen that the moderate level of room noise at the receive side masks the differences in quality which are

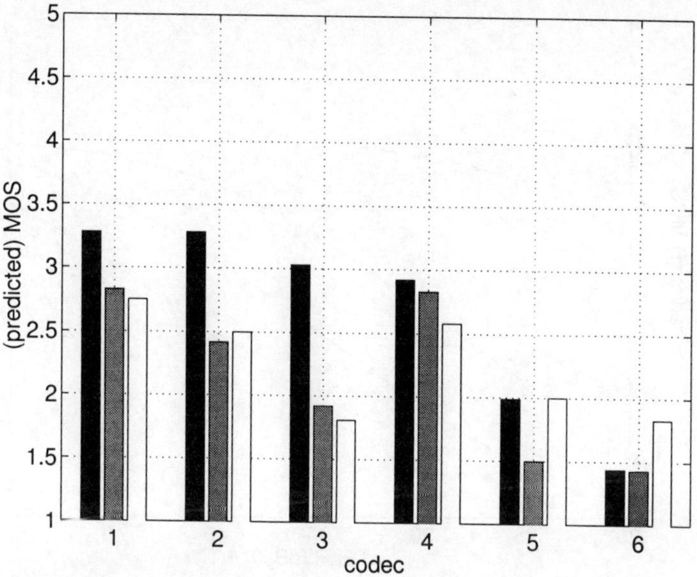

Figure 7.8. Comparison of E–model predictions (black) and SCT results regarding different codecs and $Pr = 65$ dB(A). Gray: hoth noise; white: speech babble. Test parameters: $D = 1$, $LSTR = 16$ dB, $TELR = 50$ dB, $WEPL = 80$ dB; all other parameters according to ITU–T Rec. G.107 (1998), Table 3.

normally perceived between the different codecs if otherwise 'clear' connections are tested. This is not the case for the high room noise level, although some differences between auditory test results and E–model predictions are obvious for the G.711*G.728 (codec no. 2) and the G.711*G.729 (no. 3) asynchronous tandems. However, it is impossible to decide whether the deviations for these codecs are due to a codec vs. room noise interaction.

Results of test B are reproduced in Figures 7.9 and 7.10. Once again for the moderate room noise level ($Ps = Pr = 50$ dB(A)) the subjective test results are concentrated in the center of the MOS scale. For the high room noise level ($Pr = Ps = 65$ dB(A)), E–model predictions and auditory test results match very well. Very similar results were obtained for speech babble ambient noise. The hypothesis that quality differences which are perceived for otherwise 'clear' connections may become masked by low levels of noise is supported by this test. Further support comes from the observations already described above, namely that a high noise floor had a strong influence on the quality judgments, in comparison to a nearly noise–free connection ($Nfor = -70$ dBmp). The very good correlation shown in Figure 7.10 may be astonishing with respect to the poor correlation for high noise levels observed in other tests. At least for the room noise at the receive side, comments from the subjects reveal that

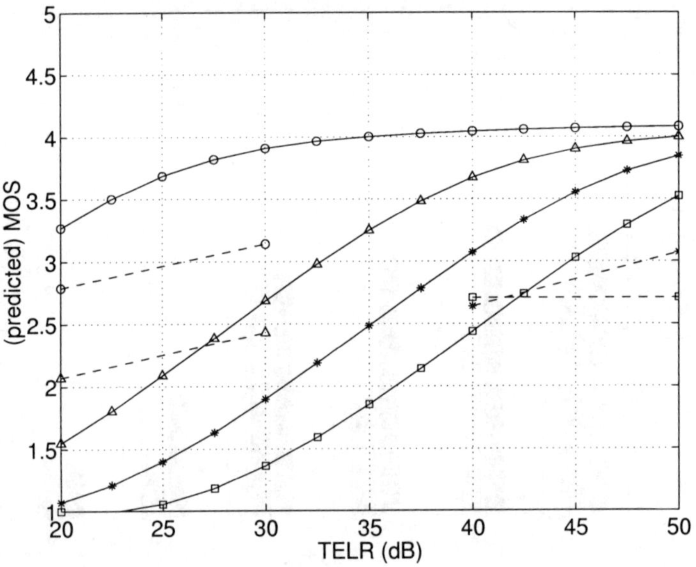

Figure 7.9. Comparison of E–model predictions (solid line) and SCT results (dashed line) for talker echo and $Ps = Pr = 50$ dB(A). o: $T = 10$ ms; Δ: $T = 50$ ms; *: $T = 100$ ms; $\square$: $T = 200$ ms. Test parameters: $D = 1$, $LSTR = 16$ dB, $WEPL = 80$ dB; all other parameters according to ITU–T Rec. G.107 (1998), Table 3.

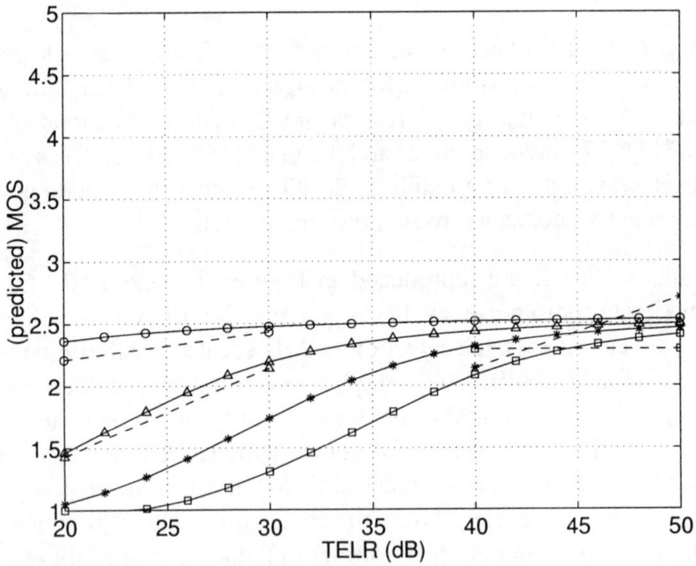

Figure 7.10. Comparison of E–model predictions (solid line) and SCT results (dashed line) for talker echo and $Ps = Pr = 65$ dB(A). o: $T = 10$ ms; Δ: $T = 50$ ms; *: $T = 100$ ms; $\square$: $T = 200$ ms. Test parameters: $D = 1$, $LSTR = 16$ dB, $WEPL = 80$ dB; all other parameters according to ITU–T Rec. G.107 (1998), Table 3.

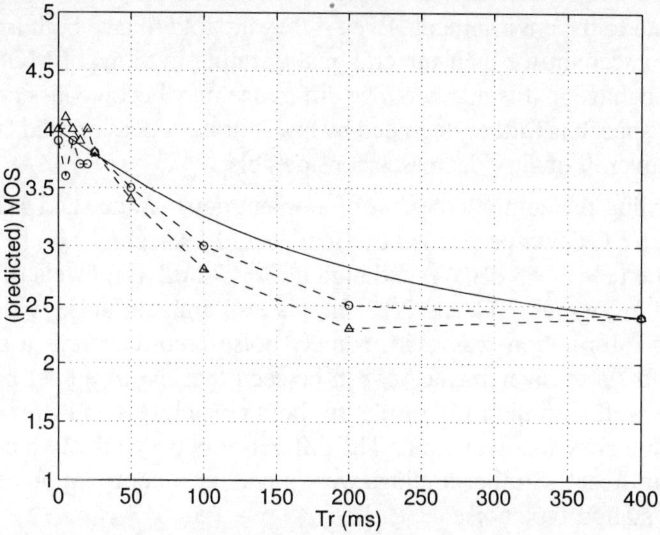

Figure 7.11. Comparison of E–model predictions (solid line) and LOT results (dashed line) for listener echo and $Ps = 48$ dB(A). o: noise from a pneumatic hammer; Δ: synthesizer music. Test parameters: $WEPL = 15$ dB, $STMR = 12$ dB, $LSTR = 15$ dB, $TELR = 50$ dB, $SLR = RLR = 10$ dB; all other parameters according to ITU–T Rec. G.107 (1998), Table 3.

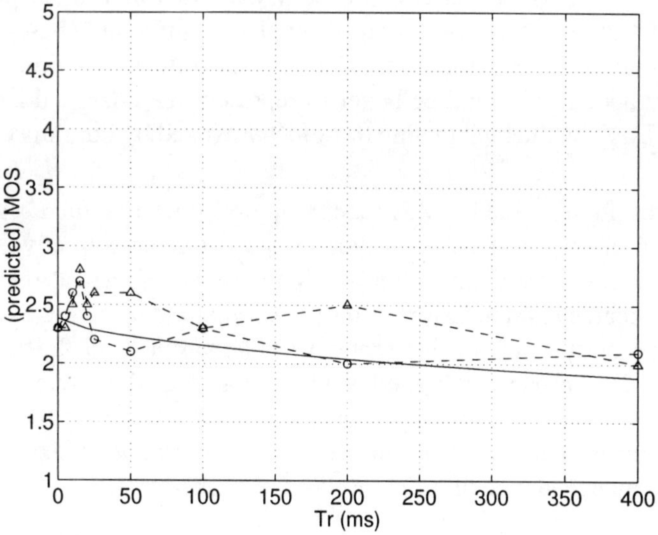

Figure 7.12. Comparison of E–model predictions (solid line) and LOT results (dashed line) for listener echo and $Ps = 72$ dB(A). o: noise from a pneumatic hammer; Δ: synthesizer music. Test parameters: $WEPL = 15$ dB, $STMR = 12$ dB, $LSTR = 15$ dB, $TELR = 50$ dB, $SLR = RLR = 10$ dB; all other parameters according to ITU–T Rec. G.107 (1998), Table 3.

they perceive a difference between impairments due to the connection and those they attribute to the environment. Even if they are told to take both impairment sources into account for their integral quality rating in terms of MOS, they do not always behave in this manner. The difference may become less perceivable if the noisy speech is further degraded by low–bitrate codecs or loud talker echo so that the overall quality decreases considerably.

Regarding the combined effect of ambient noise at the send side and listener echo, a LOT was performed by Bodden and Jekosch (1996). The results for moderate ($Ps = 48$ dB(A)) and high ($Ps = 72$ dB(A)) levels of noise are reproduced in Figures 7.11 and 7.12. In this test, ambient noise with different contents of information was used, namely noise recorded from a pneumatic hammer and synthesizer music. It can be seen that the E–model predictions agree quite well with the LOT results for both noise levels and over the whole range of listener echo attenuation. The difference between the two noise types was not significant. Similar findings were made for other room noise levels as well as for station type noise.

Further investigations into the combined effects of talker sidetone and short delay talker echoes are reported in ETSI Technical Report ETR 250 (1996, Figures 62 and 63). For this combination some masking effects can be expected because both effects are very similar in nature. It turns out that the E–model is quite effective in predicting the joint effects. The masking is taken into account in Equations G.11, G.21 and G.22 of the model algorithm, cf. Appendix G. Note that in Equation G.21 two entities completely different in nature ($TERV$ corresponding to loudness ratings vs. Ist corresponding to impairment factors) are directly added. It would be better to represent the talker sidetone in this case by $STMR$, while maintaining the effect provoked by Formula G.21 in the same way.

The results presented in this section do not prove that the E–model predicts all combinations of different types of impairments correctly. However, it shows that for some combinations which may have practical relevance the prediction accuracy is satisfactory. This is a very important result, as it does not falsify the property of adding different types of impairments on the transmission rating scale R. Probably, the good results in this regard are due to the many interrelations which occur in the respective E–model formulae for calculating the single impairment factors Is and Id. In turn, it cannot be taken as general proof of an independence of different types of impairment.

3. IMPAIRMENT FACTOR PRINCIPLE FOR LOW–BITRATE CODECS

The equipment impairment factor Ie is a new feature of the E–model in comparison to its precursors, and it makes the model particularly interesting as it

Table 7.1. Determination of a new relation $Iq = f(Q)$. Parameters are discussed in the text.

Q (dB)	no. of tests	mean MOS (tests)	E–model MOS prediction	curve parameters	MOS (fitted)	R (fitted)	Iq (fitted)
5	15	1.28	1.88		1.29	21.3	73.7
6	5	1.26	1.95		1.35	23.1	71.9
10	15	1.81	2.31		1.74	33.1	61.9
12	5	1.91	2.56		2.01	38.9	56.1
15	15	2.57	3.00	$A = 1.65$	2.49	48.4	46.4
18	5	3.01	3.51	$B = 1.71$	2.99	57.9	37.1
20	14	3.34	3.84	$C = 10.03$	3.30	53.9	31.1
24	5	3.92	4.32	$Qm = 15.96$	3.79	74.3	20.7
25	14	3.79	4.37		3.88	76.4	18.6
30	14	4.11	4.43		4.16	83.8	11.2
35	13	4.22	4.43		4.28	87.7	7.3
36	1	4.23	4.43		4.30	88.5	6.5
50	10	4.40	4.43		4.36	90.9	4.1

permits the inclusion of the effects of low–bitrate codecs. Due to their simplicity equipment impairment factors have been used as well to model the effects of imperfect echo cancellers and digital transmission errors. Nevertheless, the only values given in the model until now have been those for low–bitrate codecs.

In order to investigate the effects of low–bitrate codecs by means of Ie, it is first necessary to discuss the coverage of quantizing distortion, because the derivation of values for Ie makes use of this concept. Quantizing distortion resulting from PCM coding as well as from the MNRU is covered by the E–model formulae G.12 to G.15. A comparison of the corresponding E–model predictions with a large number of test results obtained in different laboratories reveals that the model predictions are far too optimistic, especially for low signal–to–quantizing noise ratios Q. The mean values of these results are given in Table 7.1, and the exact sources are cited by the author in ITU–T Contribution COM 12–69 (1998). In this table large differences between E–model predictions and mean MOS test results can be observed, cf. columns 3 and 4 and Figure 7.13. The mean test results have been used to determine a fitted curve according to the "equivalent Q method" using Equation 6.2. The curve is depicted in Figure 7.13, and the parameters are given in column 5 of Table 7.1. The approximated values for MOS, R and $Iq = 95 - R$ from this curve are given in columns 6–8.

The values for Iq calculated in this way are average values for a large number of laboratory tests, and thus they should coincide quite well with the E–model predictions. For this aim the curve $Iq = f(Q)$ can be modified. However, from a good matching curve it follows that even quantizing degradations which are just noticeable (30 dB $\leq Q \leq$ 40 dB) result in relatively high values for Iq (which is more realistic in listening–only than in conversational situations). A better alternative is to fit the mean test results only in the range of important quantizing distortion ($Q <$ 20 dB) and reach a steeper descent of the curve for $Q >$ 20 dB, see Figure 7.14. This behavior can be reached by replacing E–model Formulae G.12 and G.13 with

$$Iq = 15\, log_{10}\, [1 + 10^{Y} + 10^{Z}] \tag{7.2}$$

where

$$Y = \frac{Ro - 100}{15} + \frac{46}{8.4} - \frac{G}{9} \tag{7.3}$$

and

$$Z = \frac{46}{30} - \frac{G}{40} \tag{7.4}$$

G being the equivalent continuous circuit noise. The behavior of the original model as well as the modified version are depicted in Figure 7.14. Effects of quantizing distortion predicted in this way correlate much better to MOS results obtained in laboratory tests.

The determination of new equipment impairment factors follows a four-step procedure (Johannesson, 1997). First, the MOS values for each test condition and for each test laboratory are converted via MNRU values (the so–called "equivalent Q method"). Then, an rms optimization to the straight line $Ie = n \cdot K$ (n being the number of codecs of the same type in tandem) gives mean values and standard deviations for each codec. The results of step two are then taken to calculate all the combinations of codecs used in the tests. The K factors obtained are linearly transformed to maximize the correlation between predicted and subjective MOS values. In a last step the Ie factor for the unknown codec is adjusted by hand to fit into the system of known equipment impairment factors for other codecs. An independent examination of this procedure makes it clear that the codecs are adjusted in order to fit into a system of equipment impairment factors already derived. Therefore, the procedure should not be used to verify Ie values, because the adjustment will hide inconsistencies. On the other hand, the first step of the procedure, using the new formulae derived above, makes it possible to compare the effects of low–bitrate codecs on the transmission rating scale. A large number of codec test results obtained at different locations have been transformed into Ie values using Formulae 7.2 to 7.4.

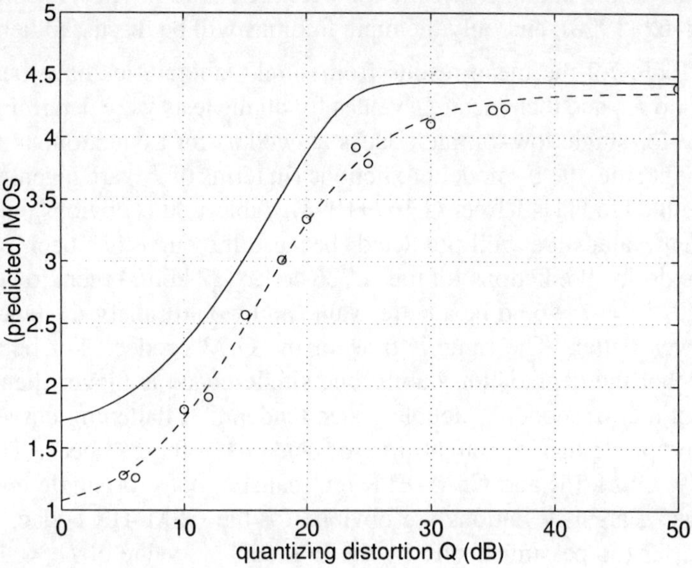

Figure 7.13. Comparison of E–model predictions (solid line) and mean test results (o) for MNRU distortions. Fitted curve (dashed line) with parameters as in Table 7.1.

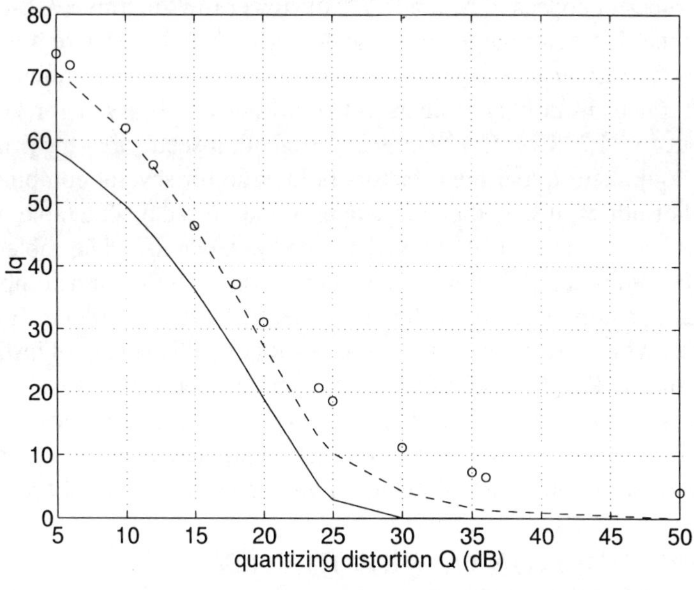

Figure 7.14. Comparison of the original E–model (solid line), the modified version using Formulae 7.2 to 7.4 (dashed line), and mean test results (o).

The details of the author's investigation can be found in ITU–T Contribution COM 12–69 (1998), and only the main findings will be discussed here.

In Table 7.2 the mean results from single laboratories have been transformed into Ie, and then mean Ie values for all the tests were determined. This was done for single low–bitrate codecs as well as for asynchronous tandems. On the right side, the E–model predictions (in terms of Ie) are given, using the values defined in ITU–T Rec. G.107 (1998), Table 1. It is obvious that MNRU impairment values are well predicted, because the curve has been purposely chosen to do so. Predictions for the G.726 codec (32 kbit/s) seem to be slightly optimistic, $Ie = 8$ would be a better value and would satisfy the tandem conditions even better. The same is true for the G.728 codec. For both codecs it seems that the degradation due to one single codec is bigger than half the amount of a corresponding double codec tandem. A flattening curve may be more appropriate than a straight line for tandems of these codecs. The values for G.729, GSM–FR and GSM–EFR are realistic, both for single and double encodings. Larger deviations are obvious for the GSM–HR codec, which is predicted far too pessimistically by the E–model. A value of $Ie = 15$ would be realistic for this codec. For the IS–54 codec the E–model is slightly too optimistic.

It is interesting to note that the equipment impairment factor principle (and the underlying additivity) does also allow realistic predictions for tandems of different types of codecs to be made, cf. the lower rows of Table 7.2. However, the differences become larger in those cases, both in a negative and positive direction. The order effect of codec tandems (i.e. that the combination A*B would be rated differently from B*A) could not be observed for G.726 and G.729, but for G.728 vs. G.729 codec. All in all, it seems that the principle of additive equipment impairment factors holds true for several combinations of codecs. For others, it would be more appropriate to establish a table of overall Ie values which have to be derived for codec tandems. The table will not respect the additivity of Ies, but gives integral values for certain combinations, as well as sometimes giving different values for reversed orders. It could be established with more auditory test data, as well as with the help of instrumental methods such as PSQM or TOSQA. Such a table can reduce the sometimes large differences between auditory test results and E–model predictions (e.g. in the last row). If network planners rely on falsely predicted Ie values, this could lead to bad voice transmission quality or an over–engineering of the network.

4. PREDICTION OF FREQUENCY CHARACTERISTICS

Some of the effects caused by linear distortion and by the bandwidth of the transmission channel have already been discussed in Section 1.3 of Chapter 3.

Table 7.2. Comparison of mean test results for low–bitrate codecs and E–model predictions. Single test results are given in ITU–T Contribution COM 12–69 (1998). *: asynchronous tandem of different codecs; nx...: asynchronous tandem of n equal codecs.

codec type	circuit condition	mean Ie (tests)	Ie (E–model)
MNRU	MNRU 5	71.4	70.7
	MNRU 10	60.2	61.0
	MNRU 20	27.8	27.3
	MNRU 35	2.5	1.6
	MNRU 50	0.6	0.0
G.726 (ADPCM)	G.726	9.1	7
	2 x G.726	19.3	14
	3 x G.726	25.2	21
	4 x G.726	31.2	28
G.728 (LD–CELP)	G.728	11.2	7
	2 x G.728	18.1	14
	3 x G.728	24.2	21
	4 x G.728	32.1	28
G.729 (CS–ACELP)	G.729	9.5	12
	3 x G.729	39.9	36
GSM–FR (RPE–LTP)	GSM–FR	18.9	20
	2 x GSM–FR	35.1	40
GSM–HR (VSELP)	GSM–HR	12.8	23
	2 x GSM–HR	31.7	46
GSM–EFR (ACELP)	GSM–EFR	1.9	3
	2 x GSM–EFR	4.8	6
IS–54 (VSELP)	IS–54	23.0	20
	2 x IS–54	45.6	40
mixed	G.729 * G.726	19.8	19
	G.726 * G.729	18.7	19
	G.729 * G.728	25.2	19
	G.728 * G.729	19.4	19
	G.729 * IS–54	31.0	32
	G.729 * G.729 * G.728	41.5	31
	GSM–FR * G.726 * G.729	29.8	39
	GSM–FR * G.729 * IS–54	38.3	52

These data, however, only give general guidance as the experimental conditions and the focus of the experiments differ between experiments. For the prediction models, it is important to analyze to what extent variations in the frequency characteristic of the transmission path lead to differences in perceived integral quality. If differences exist, network planning models should take them into account. In the first view, this seems to be easier for models using frequency characteristics as input parameters. Therefore, in this section comparisons are made both for the SUBMOD model and for the E–model.

As a part of this work, a conversation and a listening–only test were carried out, using different frequency responses for the SLR and RLR filters (corresponding to different handset characteristics in reality). The overall attenuation was adjusted in such a way that the loudness ratings calculated in accordance with ITU–T Rec. P.79 (1993) were identical. In informal auditory tests it turned out that this does not necessarily result in an equal loudness percept. Especially for strongly band–limited channels the voice is perceived as shrill. To cater for this effect three conditions were included, both with a standard attenuation and with a higher overall attenuation. All the conditions were combined either with a standard logarithmic PCM codec or with a high degree of quantizing distortion introduced by an MNRU. Thus the results allow an investigation of whether potential differences are only perceived for a clear channel, or also in the presence of other degradations. For the LOT, a slightly lower level of noise floor (-70 dBmp instead of -64 dBmp) and a slightly higher overall attenuation ($OLR = 10$ dB instead of $OLR = 5$ dB) was used. The frequency characteristics include a HP with a cut–off frequency of 1 kHz (flat above), two LP characteristics with a cut–off frequency of 2 kHz (-10 and -20 dB at 3.4 kHz), and two BP characteristics (0.5–2.5 kHz or 0.8–1.6 kHz).

For the E–model it can be seen that the transmission path characteristics have been matched regarding their loudness rating, so that the predicted values are the same for all frequency characteristics. The test results, both from listening–only and conversation tests, show that this behavior is not realistic. Especially more severe restrictions on the transmitted band result in a strong quality degradation (condition no. 5). The effect is very similar for the conversation and for the listening–only situation. A narrowing of the transmitted channel thus affects not only a listening impression, but reduces communication quality. For the connections already degraded by quantizing distortions the effect is slightly less severe, but a very similar rank order of the frequency distortion conditions can be observed (except condition no. 4). The behavior of the E–model is due to a lack of information about the frequency response, the differences in loudness rating being neutralized. The three conditions with a higher overall attenuation ($OLR = 18$ dB) yield very similar results. Thus the negative effect of a frequency distortion is not due to an excessively loud sound impression.

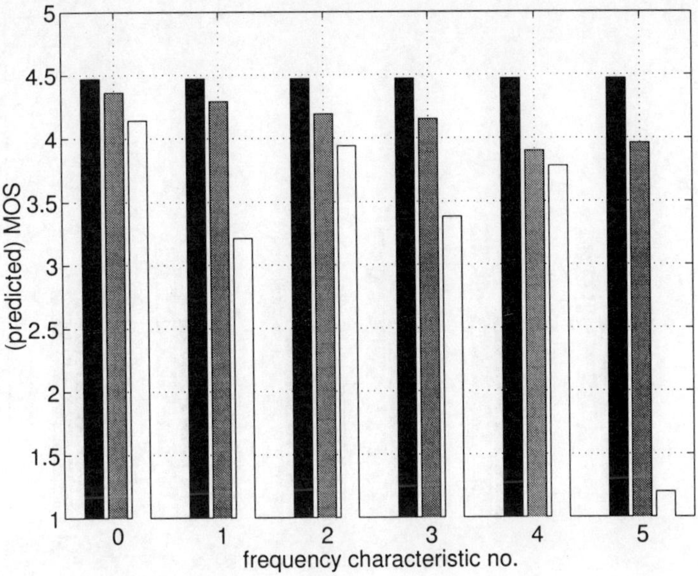

Figure 7.15. Comparison of E–model predictions (black), SUBMOD model predictions (gray) and SCT results (white), test regarding different frequency characteristics and G.711 codec. 0: standard IRS response.

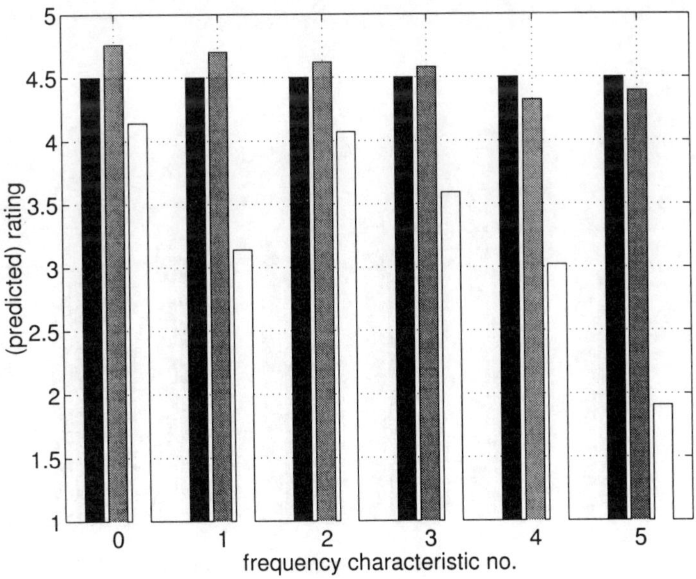

Figure 7.16. Comparison of E–model predictions (black), SUBMOD model predictions (gray) and LOT results (white), test regarding different frequency characteristics and G.711 codec. 0: standard IRS response. Ratings in terms of MOS (E–model and LOT results) and Y_{LE} (SUBMOD model).

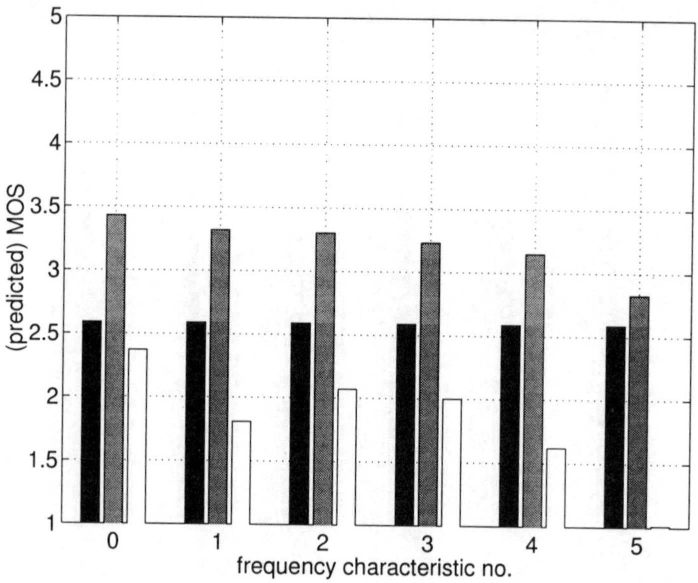

Figure 7.17. Comparison of E–model predictions (black), SUBMOD model predictions (gray) and SCT results (white), test regarding different frequency characteristics and MNRU with $Q = 15$ dB. 0: standard IRS response.

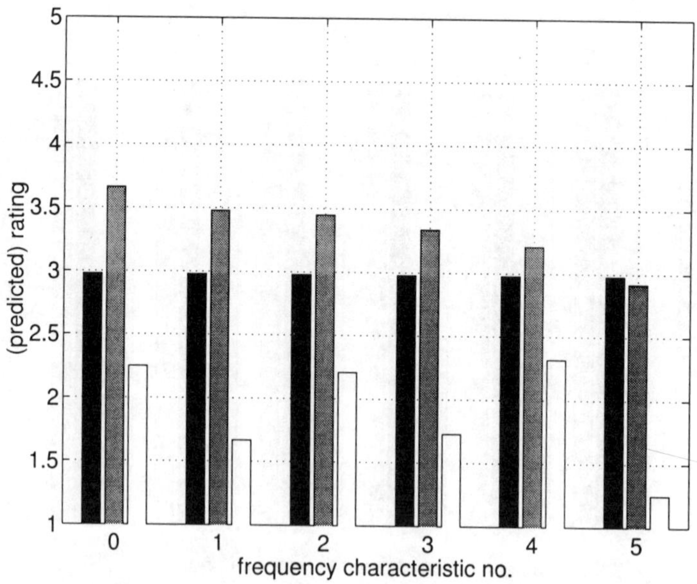

Figure 7.18. Comparison of E–model predictions (black), SUBMOD model predictions (gray) and LOT results (white), test regarding different frequency characteristics and MNRU with $Q = 15$ dB. 0: standard IRS response. Ratings in terms of MOS (E–model and LOT results) and Y_{LE} (SUBMOD model).

Unfortunately the SUBMOD model does not predict the observed effect in a satisfying way either. This is important because the model uses the frequency characteristic as an input parameter. The model's predictions show a slight effect of the transmission channel characteristic, but it is estimated too weakly by far. For the listening–only test it should be noted that the SUBMOD model gives listening–effort estimates (Y_{LE}) instead of listening–quality MOS. However, the differences in the behavior between Y_{LE} and Y_C are minimal.

The question arises as to whether the SUBMOD model can be improved to more accurately predict auditory test results for linear distortions. The experimental basis for the model will probably be restricted to transmission channels which have been found in real networks. This is not the case for the experiments described above, where the transmitted band is more severely affected. As discussed in Section 1 of Chapter 5 the listening–effort score Y_{LE} is derived from LOI. Because the relation between LOI and Y_{LE} is a monotonic one (cf. Equation 5.4), Y_{LE} becomes larger when the right side of Equation 5.3 increases. Given fixed values for A and D (fixed loudness level of received speech, constant noise level), the frequency dependent contributions to Y_{LE} consist of B'_i and $P(Z)$. These quantities are depicted in Figures 7.19 and 7.20, together with the sensation level Z.

In Figure 7.19 it can be seen that all the contributions with $Z \gtrapprox -20$ dB contribute to $P(Z)$. The frequency weighting function B' provides a weighting with emphasis on the lower frequency components; however, the weighting is approximately constant for $f \leq 1500$ Hz. In the case of the strongly band–limited transmission (800–1600 Hz, see Figure 7.20) it can be seen that though there is a strong decrease in sensation level Z, there is still a considerable contribution from $P(Z)$ even outside the transmitted band. In this way, there are strong influences on Y_{LE} from the low–frequency part as well as from the upper part of the spectrum, outside the transmitted band. This is not reasonable, because the central frequencies will dominate the overall sound perception. Apparently the growth function $P(Z)$ does not provide reasonable modeling if the variations in sensation level are too strong. These extreme cases are not covered by the model because they are generally not encountered in reality. Improvements to the model can only be made by modifying both the growth function $P(Z)$ and the weighting function B'. Another possibility is to introduce a different sound weighting algorithm independent from the contributions to LOI which are, apparently, more related to loudness and articulation. This algorithm could modify LOI and thus Y_{LE} directly, without being part of the summation in Equation 5.3.

Instrumental models for single transmission aspects (see Section 3 of Chapter 5) are able to take into account the frequency distortion of the transmission channel, by means of the input and output signals. A model which

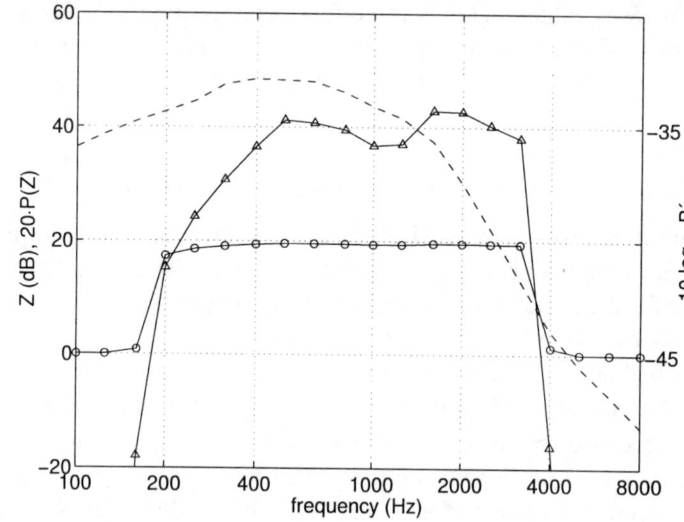

Figure 7.19. Contributions to Y_{LE} according to the SUBMOD model. Δ: sensation level Z; o: growth function $P(Z)$; dashed line: weighting function B'. Values calculated for a standard handset (IRS characteristic) used in the test.

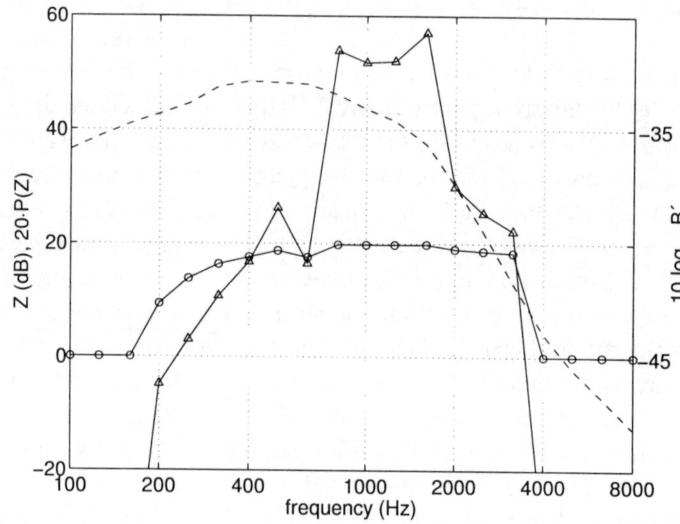

Figure 7.20. Contributions to Y_{LE} according to the SUBMOD model. Δ: sensation level Z; o: growth function $P(Z)$; dashed line: weighting function B'. Values calculated for a BP filtered transmission path (800–1600 Hz) used in the test.

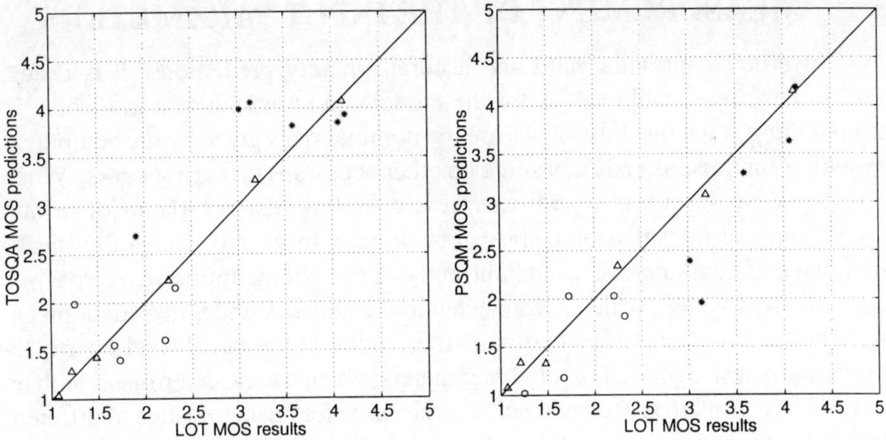

Figure 7.21. Comparison of LOT results and instrumental model predictions, test regarding frequency distortions. Left: TOSQA; right: PSQM. Δ: MNRU; ∗: frequency distortions; o: MNRU and frequency distortions.

explicitly models linear distortion is the German Telekom TOSQA model. It estimates the linear distortions and corrects for them by weighting the difference between input and output representation with this function. The model which is standardized by the ITU–T, PSQM, does not perform such a correction; therefore, linear frequency distortions have a more severe impact on PSQM predictions than on TOSQA predictions. The results from the listening–only test given above have been compared to both PSQM and TOSQA. This comparison is described in detail in Möller and Berger (1998). The correlation diagrams are reproduced in Figure 7.21. For both instrumental models correlation with LOT MOS is relatively high (0.92). It can be seen that most of the test conditions which are underestimated (i.e. rated too pessimistically) by one model are overestimated by the other, and vice versa. An ideal model for predicting the test results should make a half–way compensation of the effects of linear frequency distortion. Such a model could then, in a different form, be included in the network planning models. As the E–model does not use frequency characteristics as an input, it is not clear how linear distortions can be catered for. 'Impairment' and 'sound' seem to be two different dimensions in quality perception (cf. Section 6 of Chapter 6). Therefore it is in the first instance not appropriate to introduce another 'sound impairment factor'. A more promising way could be to establish a new sound rating model, which then could either be combined or used in parallel with an 'impairment prediction model' such as the E–model.

5. MEASUREMENT OF THE INPUT PARAMETERS

In order to provide valid and accurate quality predictions, it is necessary to have reasonable values for the model's input parameters available. In the planning stage the only values are the nominal ones given by the equipment providers, or those already measured on other occasions in real networks. When a network has been set up and when it is operating, characteristics of certain components or transmission paths can be directly measured. Generally speaking, two different approaches are currently in use. In the intrusive approach, a measurement system which is connected to the network under investigation establishes test connections and provides measurements on signal levels, partly by exchanging test signals. Potential parameters which can be determined with an intrusive system are speech and noise levels, transfer characteristics of different transmission paths, echo path loss and delay, as well as some other characteristics of the transmission path (clipping, interruptions, etc.). From the transfer functions, loudness ratings can be calculated, which in turn can be used as input parameters to e.g. the E–model.

On the other hand, non–intrusive measurement devices are available which allow in–service measurements of the signals on the transmission path to be made. These so–called 'in–service non–intrusive measurement devices' (INMDs) are primarily used to measure voice–grade parameters such as speech level, (continuous) noise level, echo loss and echo path delay. Sometimes INMDs can also provide estimates of parameters associated with digital transmission equipment (e.g. quantizing noise). When the INMD is connected as a service monitor to the network, the measurements apply to the network connection from each talker to the INMD. In general, no information can be obtained on the receive part of the connection (i.e. from the INMD to the listener), except for the echo path estimation.

Different proposals have been made to map INMD measurements on quality prediction models. For the E–model SLR can be estimated from the speech level, and an estimation for OLR can be obtained by assuming an average RLR of the network. The noise level No, referring to all the noise components (circuit noise, room noise), can be estimated from noise level measurements in both transmission directions. Together with estimated echo parameters, the E–model can be used in a simplified version, including those effects for which input parameter are available (ITU–T Delayed Contribution D.088, 1998). Until now, no estimations on e.g. the type of codec used in the connections have been documented. One possibility might be an automatic classification of speech signals according to certain characteristics introduced by the codecs, e.g. using vector quantizers or neural networks.

A different method making direct use of the non–intrusive parameters has been proposed by British Telecom, called the Call Clarity Index, CCI (ITU–

T Delayed Contribution D.088, 1998). To a large extent, it makes use of the SUBMOD model, in that a listening opinion index LOI is calculated from the sensation level, and then transformed into a conversation opinion predictor Y_C. The latter is then modified to take into account talker echo effects. The CCI has the advantage of not requiring the calculation of loudness ratings, but the quantities measured by an INMD directly form an input into the model. In ITU–T Delayed Contribution D.069 (1998) a comparison was made between auditory conversation test results and model predictions based on the INMD measurements for the test connections. A slightly lower absolute error was observed for the CCI. However, it turned out that both the CCI and the E–model falsely predict MOS for conditions including nonlinear impairments (e.g. quantizing distortion) when input parameters are measured with INMDs. Correlations between predicted MOS and those obtained in auditory tests were 0.86–0.89 for circuits with linear impairments only. On the other hand, the use of nonlinear equipment will become much more common in the future. Therefore, enhancements to intrusive as well as to non–intrusive measurements will be required in order to obtain reasonable predictions for modern network scenarios. Especially the effect of low–bitrate codecs should be covered.

6. ACCURACY OF QUALITY PREDICTIONS, LIMITATIONS

The accuracy of predictions from the network planning models can only be determined by comparing them with laboratory test results and experience gained when using the models. Experience of using the model to investigate different transmission scenarios is reported elsewhere, see e.g. ITU–T Contribution COM 12–76 (1996); ITU–T Delayed Contribution D.093 (1998). A comparison with LOT or CT results is mainly made in terms of MOS. However, the E–model performs its primary prediction on the transmission rating scale. Differences to test results may therefore be due to false model predictions in terms of R, or to a bad transformation from R to MOS (or both). A spread in the results obtained under various test conditions cannot be balanced out by normalization or transformation, see Section 2 of Chapter 6. Thus, many sources of imprecision have an influence on the comparison.

Even if precise MOS results can be obtained for a specific scenario, the standard deviation for ratings on the MOS scale is about 0.7 categories (value given in ETSI Technical Report ETR 250 (1996), p. 165). In some experiments carried out here even larger standard deviations were observed, especially for room noise conditions. The upper end of the MOS scale is very rarely used for rating telephone transmission quality, a fact which is already taken into account by the E–model (predicted MOS $\in [1; 4.5]$). Transferred to the R–scale, this spread easily results in differences of up to 20. Therefore, it is not appropriate to verify model predictions by performing a statistical analysis regarding the significance of differences for observed results. Instead users of such models should look at all the information available to them from all potential sources, and especially at experiences gained from similar problems. Even when model predictions lead to a result about 10 points above a "limiting threshold" (see Table 6.1) it cannot be guaranteed that users of such a system will be satisfied. A security margin that is as high as possible should be applied when using model predictions to plan future telecommunication networks.

7. SUMMARY

It has become obvious that a multitude of different factors has an influence on the quality of a service, and that, in general, auditory testing is required to measure quality. Nevertheless it is important to have planning tools at hand which can provide reasonable quality estimates for specific domains and network configurations. The quality predictions in terms of MOS scores from two different models were investigated by comparing them to auditory test data. Most of the tests were performed during the work on this book. The focus of the analysis was on the E–model, whereas the SUBMOD model is more interesting for the prediction of frequency–dependent features, such as the frequency characteristic of the transmission channel.

For speech communication systems which are degraded by one specific perceptive characteristic or impairment it turns out that E–model predictions are realistic for several degradations occurring in traditional networks, e.g. talker echo and listener echo. The effects of room noise are not appropriately covered by the E–model, because it disregards the Lombard effect. A modification was proposed in this work in order to enhance the model, based on the comparison with auditory test data. The effects of long overall delay are predicted in a way that emphasizes highly interactive conversations. For normal, free conversation, the model's predictions seem to be too pessimistic.

The equipment impairment factor is a new feature of the E–model (compared to its precursors), and it renders the model particularly interesting for estimating the perceptive effects of low–bitrate codecs. The methodology to derive such factors makes use of an indirect comparison with model predictions for quantizing distortion. Unfortunately, the model is very imprecise in predicting the perceptive effects of quantizing distortion. Based on auditory test data the according E–model formulae were modified, leading to a better prediction for such impairments as well as to reasonable equipment impairment factors. The analysis shows that the equipment impairment factor methodology provides realistic estimates of degradations introduced by several low–bitrate codecs. However, limitations exist for asynchronous tandems of such codecs.

For that purpose, a table derived with the help of the instrumental models discussed in Section 3 of Chapter 5 would be more appropriate.

A strong advantage of transmission planning models is that they can predict the joint effects of different degradations occurring simultaneously. Namely the E–model regards a multitude of impairments, whose perceptive effects shall be additive on the transmission rating scale. A comparison with auditory tests carried out in this book illustrates that combinations of low–bitrate codecs and room noise as well as talker echo and noise can lead to a masking of one impairment by another, a fact which is not always realistically covered by the E–model. The effect can be observed especially for moderate levels of room noise. Nevertheless, the comparison does not falsify the underlying additivity principle of the E–model in a general way. The relatively realistic prediction results for combinations of different impairments are due to the many interrelationships which occur in the according E–model formulae for calculating single impairment factors.

Both the SUBMOD as well as the E–model do not yield realistic quality estimations when the frequency characteristic of the transmission channel deviates from standard characteristics. Apparently, the loudness rating methodology used in the E–model is not adequate in measuring impact on the color of sound. Though using explicit frequency information as an input, the SUBMOD model makes use of a very similar principle, as it was discussed in Section 1 of Chapter 5. Thus, the effects of linear distortions are not covered by both transmission rating models. Predictions of other instrumental models (e.g. PSQM or TOSQA) may provide additional information regarding the effects of low–bitrate codec tandems and linear distortions. An integration of different modeling approaches promises better quality predictions in those cases. Further combinations are necessary when the input parameters of the models have to be measured in real–life networks. Provided that realistic quality predictions can be obtained from those combinations, the existing approaches form a basis for new quality monitoring models to be developed in the future.

Chapter 8

FINAL DISCUSSION AND CONCLUSIONS

The quality of telephone speech is a multidimensional property. It arises when certain perceivable characteristics of the speech itself meet the users expectations and desires. Speech as such does not carry a quality in itself. Quality is an event, and consequently it can only be established in a specific situation. A schematic describing the constitutional components of the quality of a telecommunication service has been developed. This schematic is meant to be user directed, in the sense that different dimensions of quality are related to concepts of usability, utility or acceptability. The latter describe at which different levels the performance of a telecommunication service is affected in the eyes (ears) of the user. The entities of the schematic have been put into relation with characteristics of the transmission system that have an impact on certain quality features. To gain a better understanding of the perceptive factors which result in percepts of loudness, articulation or disturbance during the conversation, the psychoacoustic background has been presented. This background is also the basis of one of the transmission rating models discussed later on.

Quality assessment is a process which implies the comparison of a perceived entity with an internal reference. In telephony different types of references exist, depending on the experience of the user, on which types of calls (local, international) are mostly made, on which type of transmission (wired, mobile) and terminal equipment are used. Results of quality assessment tasks reflect different experience horizons of a limited and mostly non–representative test group. Therefore, results from quality assessment experiments cannot normally be generalized in terms of what can be expected by a specific, representative user group. In contrast, assessment scores reflect to a high degree the situation in which they have been gathered. Especially in telephony, laboratory tests are most often limited to the listening–only situation, because of time and money constraints. Such tests differ significantly from the situation which a user of a telecommunication system encounters. Proposals have been made in this work for a more economic type of conversation test. The SCT scenarios presented here enable more tests to be carried out in realistic dialog situations. It has been shown that it is very difficult to generalize test results to conform to a more 'absolute' level, including transformation, as is the current practice in telephony. Test results obtained in one specific situation should therefore always be seen in this situative and subjective context. Nonetheless, it is possible to identify inter–individually valid quality features, which are necessary in order to plan quality for a larger number of users. Most often quality assessment has exactly the task of measuring inter–individually valid properties. A large number of assessment methods have been described and related to features of speech communication quality and usability.

A certain degree of generalization is thus necessary for a network planner so that he can choose network configurations leading to satisfactory speech communication quality for the largest possible user group. Transmission rating models are an adequate tool for predicting a mean of general subjective ratings on speech communication quality, in a precision and validity which seems to be sufficient for the planning of telecommunication networks. It should be emphasized that network planning models do not aim to predict the quality of service or speech communication quality, as perceived by one specific conversation partner in a specific situation. This may differ to a considerable degree from what models predict. It is especially dangerous to directly compare model predictions for different types of impairment on an 'absolute' level (e.g. a certain level of room noise would be equivalent in quality degradation to a specific talker echo). Model predictions should rather be seen as relative estimates for comparative planning purposes. Good model prediction validity will not be found for the configurations which were not taken into account when the model was established. Improvements and extensions for such models were discussed. The limitations of the current models will have a major impact on their usefulness for planning future network scenarios. It is therefore necessary to extend existing models or to produce new ones in order to avoid misuse outside the scope and outside the sector of model validity.

A specific class of quality prediction models is devoted to the planning of telecommunication networks. These models aim at producing quality estimates on different scales based on instrumentally measurable and nominal characteristics of the network configuration. Two network planning tools were analyzed in more detail regarding the validity of their respective predictions. It turned out that the E–model produces valid estimations for a variety of impairments encountered in existing networks. For other degradations its predictive power is limited. Enhancements to the E–model have been proposed so that the effects of ambient noise (Lombard effect) and quantizing distortion can be included better. The latter modification also enhances the methodology for deriving new equipment impairment factors. Values for Ie were shown to partly correspond to auditory test results. Other Ie values will have to be updated in order to achieve more realistic predictions. The coverage of overall delay was shown to be limited to highly interactive tasks, or experienced users.

One advantage of the E–model is that it takes into account a multitude of different types of impairment, by treating them to a certain degree independently on an imaginary "psychological scale". This principle has been analyzed in theory and by performing auditory experiments. It turns out that for some combinations of impairments the model predictions are reasonable. In other cases, e.g. for moderate levels of noise, masking occurs which makes the model predictions too pessimistic. Several masking effects are already included in the E–model, via the interrelationships in calculating the respective impairment factors. The additivity is also questionable for tandems of several low–bitrate codecs. Such combinations would be more appropriately covered by setting up tables with overall impairment factors for the complete tandems. The instrumental models discussed in Section 3 of Chapter 5 will be a valuable tool for establishing such a table.

Whereas the E–model does make use of frequency–weighted scalar parameter values for describing the single transmission paths, the SUBMOD model explicitly requires frequency characteristics. However, both approaches do not lead to realistic predictions when severe limitations of the transmitted frequency band are introduced. The reason for this is that both prediction models are very similar to the model used for calculating loudness. Differences in quality ratings for transmission channels resulting in similar loudness but different sound impression can therefore not be predicted. Instrumental models (Section 3 of Chapter 5) promise a better prediction of such sound–related degradations. It seems worth establishing a sound–related quality model which aims at predicting a different quality dimension, starting off from the frequency characteristic of the transmission path as an input value. The model could make use of the coverage of linear distortions in TOSQA, however not by neutralizing them.

The E–model is a distillate of several approaches for predicting quality as experienced by an average user and based on instrumentally measurable input parameters. With the inclusion of new types of degradations (e.g. those due to low–bitrate codecs) this principle was abandoned in favor of integral equipment impairment factors. The model will only remain useful if it is continually updated to include most of the degradations introduced by modern equipment. It seems neither possible nor advantageous to establish new equipment impairment factors for each and every piece of equipment. To maintain the model's basic principle, a classification of degradations occurring in modern networks has been proposed (cf. Appendix B). This classification can help to reduce the somehow complex types of impairment to more elementary entities. It would serve to make the model more manageable, so that it relies less on single, specific pieces of equipment. The model's predictive power for new equipment showing similar degradation to known pieces of equipment will increase in this way.

In its present state the E–model is able to make predictions on narrow-band network scenarios including wirebound as well as wireless communication systems (see Figure 8.1, case a). However, many degradations which occur only in mobile scenarios are not yet covered by the model. Examples include clipping and interruption in the transmission, impulsive noise, noise contrast, etc. In the next stage the model could be updated in order to include such effects (case b). Other degradations occur in packet based systems, e.g. frame or packet loss, cell delay variation effects, etc. The evolving market in terminal equipment will make it necessary to include e.g. headphones and hands–free terminals in the network planning models. This is not obvious as the send and receive loudness ratings do not seem to be adequate, and especially with HFTs effects of the talking and listening room will gain importance. Additional investigations are necessary to clarify the internal reference (quality horizon) under which users regard telecommunication systems not operated via traditional handset telephones (case c).

When transmission bandwidth is less of a problem, wideband (100–7000 Hz) systems play a major role. Such systems use either ISDN or packet based systems for transmission, and have other terminal equipment, e.g. headphones or hands–free terminals. All the traditional transmission planning models, including the E–model, have been developed for narrow–band speech, so their predicting power will not cover wideband transmission paths. Particularly in the investigation into linear distortions it turned out that transmission channels differing in bandwidth from traditional ones are not covered by the model predictions. Instead of extending existing models to wideband scenarios it seems more appropriate to develop completely new models, on the basis of new auditory (conversation) test results (case d in Figure 8.1). Fundamental principles of narrow–band models can, of course, continue to be used. Because of the lack of test results it cannot be decided whether wideband loudness ratings (ITU–T Contribution COM 12–15, 1997) should be used as input parameters for potential new models. Special attention must be paid when comparisons of narrow–band and wideband systems are made within one test session. It is obvious that the quality reference for narrow–band telephone speech will be different from the one for wideband speech, because the latter is generally not used in telephone connections. Potential users may expect e.g. a high fidelity voice transmission. In the wideband transmission scenario the quality dimension 'sound' will probably have a stronger influence on the integral speech communication quality.

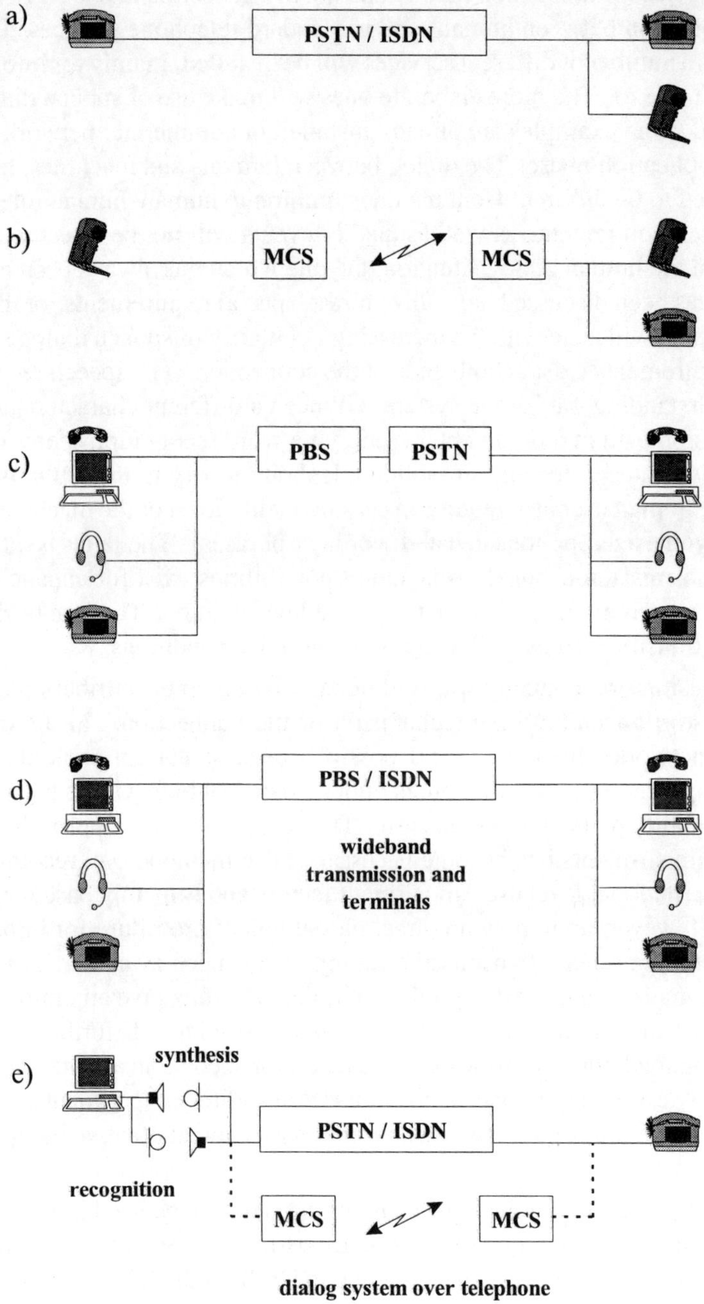

Figure 8.1. Future telecommunication scenarios.

Speech communication quality and quality of service in this work relate to communication between humans and to standard telephone services. In future networks a number of different services will be installed, mainly for information retrieval (case e). The more elaborate ones will make use of spoken dialog systems, and some examples are already installed in commercial networks, in the test or application phase. The dialog between humans and machines, however, has proven to be different from the one common to human–human interaction. Therefore, requirements as well as quality events will not be directly transferable from the human dialog situation, the one which has always been assumed in what has been discussed here. In contrast, special requirements for the transmission path will arise with the increasing popularity of spoken dialog systems. New requirements exist at both ends of the connection. The speech recognition and understanding part of the system will need a different characteristic of the transmission path in order to obtain satisfying word recognition scores (the low recognition rates especially in mobile telephony testify to this). On the other hand, the human user also requires a certain quality level of the machine speech (either synthesized or concatenated words or phrases). The latter is a function of the transmission channel. Adaptation possibilities exist for enhancing synthesized speech in order to fit the telephone line situation. This can be done for both the transmission as well as for ambient noise conditions.

Transmission planning was traditionally based on the attribution of quantizing distortion units to particular parts of the connection. In the evolving modern networks this is no longer possible, because certain parts of the connection may require higher amounts of degradation which will be balanced out by high quality parts of the connection. Therefore, an overall approach to transmission impairments has become necessary. The methodology recommended on an international level uses impairment factors and is mainly based on the E–model. However, up to now no direct measurement procedure for impairment factors and the related transmission ratings R has been available. A proposal has been made in this work, and the preliminary results give an optimistic impression of the validity and reliability of that procedure. If further validation can be obtained, the CR–10 scale could be established as an alternative method for directly estimating impairment factors from auditory experiments. This will facilitate the inclusion of new low–bitrate codecs and alternative equipment in the E–model.

Predictions from a network planning model may be used in the future to provide automatic voice quality control in existing networks. Some of the benefits of such control systems are mentioned in Draft ETSI Guide EG/STQ-00004 (1998). They include e.g. an optimization of the transmission quality during the call set–up, the possibility for the user to choose between different levels of speech communication quality, e.g. on a per–call basis including different charging according to the required quality, or simply network monitoring. To realize this, measurement capabilities for the more important input parameters of the model are needed. INMD systems (cf. Section 5 of Chapter 7) seem to be a good choice for this task, as they allow measurements to be performed during the system's operation. On the other hand, intrusive systems establishing test calls are also capable of providing adequate estimations of the necessary transmission characteristics, sometimes more accurately (because special test signals can be used). The measured characteristics then have to be mapped to model input parameters. At this point information loss (e.g. by reducing frequency characteristics to scalar loudness ratings) should be avoided. In corporate networks, operated by several providers, especially the additivity principle of the E–model is advantageous, partly because it permits impairment factors for single parts of the transmission path to be calculated (other transmission characteristics, e.g. delay times, will have to be added on an overall basis). In a final state quality prediction models may be linked to the whole product circle of planning, designing, setting up, operating and controlling high–quality telecommunication networks.

Appendix A
Glossary

Quality	Result of a judgment of a perceived constitution of an entity with regard to the desired constitution (Jekosch, 1998).
Speech Quality	Quality of spoken language when acoustically realized. Result of a perception and assessment process in which the assessing subject establishes a relation between the perceived (i.e. the auditory event) and the desired or expected.
Quality of a Service (QoS)	The collective effect of service performance which determines the degree of satisfaction of a user of a service (ITU–T Rec. E.800, 1994).
Speech Communication Quality	Speech quality as experienced in a bi– or multi–directional conversation. The communication can be either direct or indirect via a communication system.
Voice Transmission Quality	Quality of spoken language when acoustically realized and transmitted over a communication system, experienced by a subject in a listening–only situation.
Ease of Communication	Amount of effort and resources expended in order to follow a conversation. Depends on the conversation partners.
Effectiveness	Accuracy and completeness with which specified users can achieve specified goals in particular environments (ETSI Technical Report ETR 095, 1993).
Efficiency	The effort and resources expended in relation to the accuracy and completeness with which specified users can achieve specified goals (definition similar to ETSI Technical Report ETR 095 (1993)).

Conversation Effectiveness Ability of a communication system or service to provide an accurate and complete conversation for a user. Results from the system's conversational capabilities.

Communication Efficiency Amount of effort and resources expended in relation to the accuracy and completeness with which a conversation can be carried out over a communication system.

Service Performance Accuracy and completeness with which a specified person can use a telecommunication service. Depends on service support, service security, and on how well the service is adapted to a successful and easy application by a user.

Comfort The users' opinion on how much they like to work with a system, resulting from system and service capabilities as well as user attitude and emotions.

Usability Suitability of a system or service to fulfill the user's requirements. Includes effectiveness and efficiency of the system and results in user satisfaction.

Utility Usability of a system or service in relation to the financial costs (ETSI Technical Report ETR 095, 1993).

Acceptability of a Service (AoS) Multidimensional property of a service, describing how readily a customer will use the service. Represented as the ratio of the number of potential users to the quantity of the target group (EURESCOM Project P.807 Deliverable 1, 1998).

Appendix B
Perceptive Characteristics Resulting from New Technologies

With the transition from analog to digital networks, there was a clear decrease in the overall sum of transmission impairments, cf. Figure 2 of Draft ETSI Guide EG/STQ-00004 (1998). In particular loss variation, noise and analog distortions were reduced considerably. New impairments arose like nonlinear distortions (codecs) and delay, but the overall quality could be improved by introducing digital transmission, switching and terminal equipment. In the future, however, mobile and packet based systems, delay and low–bitrate codecs will have the highest impact on the quality of the speech transmission. The overall amount of impairment is expected to increase. New types of equipment and new telecommunication services will show special transmission properties which can cause voice transmission or speech communication quality to deteriorate in a way different from what occurs in traditional networks. The most commonly used pieces of this equipment are actually packet switching techniques, such as asynchronous transfer mode (ATM) or voice over internet protocol (VoIP), echo control devices like echo suppressors or echo cancellers (ECs), mobile communication systems like cordless phones (e.g. DECT) or mobile phones (e.g. GSM), hands–free terminals (HFTs), and digital circuit multiplexing equipment (DCME). The impact of these technologies is discussed in the following.

1. IMPACT OF NEW TECHNOLOGIES AND EQUIPMENT

The two major degradations caused by voice packetizing systems are overall delay and the effects of low–bitrate codecs. Overall delay in ATM systems results from packetization, the compensation of cell delay variation, switching and exchange, and the propagation itself. Packetization delay is largely dependent on the type of codec used. Kliche et al. (1997) give values for different codecs standardized by the ITU. For a G.711 codec (logarithmic PCM) e.g., this results in a typical overall delay of 16.5 ms (ETSI Technical Report ETR 250, 1996, p. 87). In voice transmission over internet protocol, much higher delay times are to be expected. In Draft ETSI Guide EG/STQ-00004 (1998) three different classes of VoIP terminals are defined with maximal overall delay times of 150 ms, 250 ms and 450 ms. High delay values also require a high echo return loss at the opposite terminal, or the use of echo control devices. An additional problem in packetized switching techniques is cell or packet loss, and cell delay variation. Cell loss occurs when the data traffic is high and the packets arrive so late that they cannot be used for the reconstruction of the speech signal. The ratio of lost packets depends on the traffic load as well as on the length of the FIFO receiver buffer. Whereas the traffic load can be reduced by providing sufficient channel capacity, the receiver buffer should not be too long so as to limit the additional delay that it introduces. If the traffic load exceeds the channel capacity, this results in interruptions in the speech link. As the FIFO buffer will not always be of fixed length, cell delay variation can cause further impairments, namely that the speech signal cannot be reconstructed accurately at packet boundaries in the case of buffer length variation. VoIP terminals are often different from normal telephone handsets, and the acoustic characteristics of the terminals may be a further source of degradation. The call set–up time is an additional parameter which influences the quality of service especially of VoIP systems.

Mobile phones and the associated transmission techniques introduce impairments which are not encountered in wirebound telephony. To a lesser extent, the same is true for cordless phones designed for within–building communication. One of the main impairments is due to the codec used. Especially in tandems (i.e. cascades of codecs) the degradation will be important, and some results are discussed in Section 3 of Chapter 7. The impact of transmission errors, which occur either randomly or due to the erasure of complete frames, depends on the particular codec used. Signal fluctuations due to multiple propagation paths as well as interference from other users cause degradations which are now regarded as ‘typical’ for mobile scenarios. Also speech clipping, due to the insertion of voice activity detection devices, produces an impairment which users of the wirebound telephone are unfamiliar with so far. In conjunction with background noise, the voice activity detection (VAD) causes an effect called “noise contrast”, i.e. differences in background noise dependent on whether speech is present or not. To avoid this effect, some devices introduce a so–called comfort noise. However, this artificially introduced noise is normally different in spectrum and temporal characteristics from background noise transmitted over a telephone line. Sometimes a filtering with the long–term spectrum of the background noise is applied to reduce this contrast.

In addition to these mobile–specific impairments mobile communication systems contribute to the overall delay and talker echo. The acoustic characteristics of mobile phones (in terms of the coupling to the ear, and the D–factor) are generally worse than those of conventional handsets (Gierlich, 1996). The

D–factor is particularly poor due to the bad position of the microphone in relation to the talker's mouth. The acoustic properties are also a problem for the hands–free terminals. In general, the talker is located at some distance from the microphone, and the remote loudspeaker does not provide a quality comparable to that of a traditional handset. The acoustic room characteristics lead to reverberation, and the acoustic echo has to be reduced by specific circuits. These circuits can introduce speech clipping in return, or (in the worst case) permit only half–duplex connections, i.e. conversation is only possible in one direction at a time.

Echo canceller and echo suppressor devices try to reduce talker echo by suppressing or compensating signals coupled in the return path, aiming at avoiding degradation of the transmitted speech quality. The more elaborate principle is the echo canceller (EC), where potential echo is estimated and then subtracted from the send path. The estimation algorithms need some convergence time to produce valid estimates, and interruptions in the receive path may occur. Especially when non–linear equipment is included in the echo path, the echo estimation will not be perfect, and so further degradation of the speech signal is the result. Residual echo remains because no perfect cancellation can be achieved. Echo control is always necessary if the total transmission time increases and signal reflections occur. Delay is caused e.g. by long distance satellite connections and mobile connections. Another element which introduces delay is digital circuit multiplication equipment, DCME. It serves the purpose of time–domain multiplexing and makes use of medium–rate codecs, e.g. ITU–T G.726 or G.728 codecs. Voice activity detection and speech interpolation are further sources of degradation in this type of equipment.

2. CLASSIFICATION FOR MODELING PURPOSES

The list of equipment which leads to special perceptive characteristics of the transmitted speech signal is long and will have to be continually extended in the future. Therefore, it is not reasonable to discuss each perceptive effect in the light of the responsible equipment. Instead, a better way is to classify the types of degradation in terms of their related perceptive characteristics. In this way, future equipment can also be characterized by the types of impairment they introduce, and an estimation of the quality degradation can be given by reference to equipment with similar effects. The effects can later be included in extensions of the transmission planning models, e.g. the E-model. In Table B.1 a first classification approach is done for the equipment described above, based on the information which can be found in literature. The table is not meant to give a complete description of all the related effects, but rather it cites the more important ones and classifies them under broader problem sections. The first group corresponds to time–variant degradations, mainly due to non–continuous

Table B.1. Impact of special equipment or systems on speech communication quality.

problem	source	EC	HFT	DCME	mobile comm.	ATM	VoIP
interruptions and time variation	interruptions in the receive path (VAD)	X	X	X	X		X
	noise contrast (VAD, comfort noise)	X	X	X	X		X
	noise 'pumping'	X					
	speech clipping	X	X	X	X		X
transmission errors	random bit errors				X		
	burst errors, frame erasures				X	X	X
	signal fluctuations, fading				X		
speech degradation	coding and decoding			X	X	X	X
	cell delay variation, speech interpolation			X		X	X
	acoustic terminal characteristics		X		X		X
	ambient noise		X		X		
	interference from other users				X		
delay and talker echo	codecs, packetization			X	X	X	X
	speech degradation due to nonlinearities or acoustic echo	X	X				
	residual echo	X					
	transient echo (convergence)	X					
call set–up time							X

transmission and voice–activity detection. The time–varying behavior has some consequences for the according impact on speech communication quality, and for the assessment method to be used for investigation. The second and third group subsume constant or time–varying degradations which are related to transmission errors or which particularly affect speech, all in the listening–only situation. The impact of the fourth group regards talking and listening situations, because echo and delay are involved. The fifth problem group does not relate to the transmission phase, but to the call–setup phase, and is therefore more related to usability concepts.

Appendix C
Discussion of Articulation and Intelligibility Test Methodologies

Articulation and intelligibility tests can be divided into two categories. In the first one are the open response tests. Identification scores can be determined for different types of speech elements: sound or logatom articulation, word intelligibility, and sentence intelligibility. They are generally carried out using trained test teams ($\sim$30 training sessions). The speech material used should be adopted from the specific language, e.g. the structure of syllables, frequency spectrum/amplitude distribution, and phonetic properties. Sentence material is less preferred because intelligibility depends on intonation, listener's knowledge of grammar, sense of the phrase, etc. The subjects' task is to write down what they have heard. Manual or automatic analysis of the transcripts yields articulation or intelligibility scores which are specific for the test method and for the test vocabulary, language, etc. Relations can be established between different test methods, e.g. sentence intel. = f(word intel.), word intel. = f(sound art.), word intel. = f(logatom art.), but these relations will depend on the language used in the tests. By employing a reference system (like SRAEN, formerly defined by the ITU) an equivalent attenuation difference between the system under test and the reference system at a fixed value of e.g. 80 % sound articulation can be determined (ITU–T Handbook on Telephonometry, 1992, p. 108). This value is called the 'articulation reference equivalent' value (AEN value).

Open response tests are generally more sensitive than multiple choice tests giving a closed set of response possibilities. In this second category the rhyme tests are used most frequently. They consist in a comparison of similar sounding words, in order to get information about the underlying transmission circuits. Original work has been done by Fairbanks (1958) and Black (1957). The variants used in telephony are the Diagnostic Rhyme Test (DRT) (Voiers et al., 1975) and the Modified Rhyme Test (MRT) (House et al., 1965). In DRT, stimuli consist of CVC items (consonant–vowel–consonant). The medium and the final consonant are not examined, only the identifiability of the first consonant is tested. Subjects are asked to make a choice between two meaningful items which have a minimal phonemic contrast (minimal pair). The MRT also uses CVC items, but gives six response alternatives. The identifiability of both initial and final cluster is tested, but never simultaneously. A German version of the MRT has been developed by Sotscheck (1982). The use of meaningful test items has the advantage that no previous training is required for the test subjects. However, a major drawback is that test items are restricted, which results in little diagnostic value (Gibbon et al., 1997). Other test methodologies, e.g. the Cluster Identification Test (CLID) (Jekosch, 1992) overcome this limitation by using mainly meaningless clusters (thus an articulation test methodology), and by providing open response possibilities. However, the CLID test methodology has been verified for synthesized speech so far, so its applicability to telephone speech still has to be proven. Comparison between articulation scores obtained in open response tests and those from rhyme tests generally result in the selection method (multiple choice) showing higher scores, so ceiling effects can be observed for highly intelligible speech such as in telephone applications.

A remarkable method which aims at measuring speech quality by indirectly measuring intelligibility is the so–called speech interference test proposed by Nakatani and Dukes (1973). In a subjective LOT, speech intelligibility is measured against a background of interfering speech. The quality difference between the degraded test sample and a reference speech sample is determined as the difference in interfering speech level, namely the one for which both (test and reference samples) obtain 50 % sentence intelligibility on semantically unpredictable sentences. The test was used for discriminating highly intelligible speech samples degraded by LP and HP filtering, additive noise and telephone line transmission. The threshold difference measures relate monotonically to subjective "acceptability" ratings obtained as a linear combination of the ratings on four different scales. It should be noted, however, that this method is based on intelligibility of nonsense sentences under interfering speech conditions. Therefore, the dimension of quality is a very specific and restricted one. The interference test was set up as a two or three channel test with diotical presentation, a situation which is totally different from normal telephone situations (perhaps it is more similar to making phone calls in ambient noise situations, or when using a hands–free terminal). No evidence can be given that this methodology yields a more general measure of telephone speech quality for various telephone channel degradations.

In conversation tests intelligibility scores can also be obtained. One possibility is the analysis of the repetition rate in real–life conversations. This method is highly natural because it uses a real–life condition. However, it is not really adequate for ordinary telephone circuit quality, as it shows poor resolution and a high variance between conversations and test subjects. Other conversation methods used so far are a measure of the time to communicate an idea, or to repeat a sentence which is read by the experimenter. More details can be found in the ITU–T Handbook on Telephonometry (1992).

Appendix D
Graeco–Latin–Square Test Design

In a GLS test design four factors can be investigated, namely circuit conditions, conversation scenarios, order within the experiment and subject pairs. By using hyper–Graeco–Latin squares additional features can be catered for, e.g. test cabinet, etc. The rows of a GLS designate pairs of subjects, columns the order within the experiment run; the first alphabet designates circuit condition, and the second alphabet the conversation scenario. All four factors are equilibrated within themselves, but not all the interactions between factors are balanced out (a fact which is regarded as negligible in this context). GLSs can be constructed by combining two orthogonal Latin squares (Bortz, 1993). Different construction principles have to be used, depending on the dimensionality, see Raghavarao (1971). The resulting test plan contains n^2 combinations, and thus n^2 test condition combinations. As an example: 12 test conditions are combined with 12 different subject pairs, guaranteeing that each condition has a different order within the experiment and uses a different conversation scenario. The resulting test plan is given in Table D.1.

Table D.1. Example of a Graeco–Latin square design. Roman numbers: pairs of subjects; Arabic numbers: order within the experiment run; Latin alphabet: circuit condition; Greek alphabet: conversation scenario.

	1	2	3	4	5	6	7	8	9	10	11	12
I	Aα	Bμ	Cζ	Dη	Iε	Jδ	Kκ	Lλ	Eι	Fϑ	Gβ	Hγ
II	Bβ	Aλ	Dε	Cϑ	Jζ	Iγ	Lι	Kμ	Fκ	Eη	Hα	Gδ
III	Cγ	Dκ	Aϑ	Bε	Kη	Lβ	Iμ	Jι	Gλ	Hζ	Eδ	Fα
IV	Dδ	Cι	Bη	Aζ	Lϑ	Kα	Jλ	Iκ	Hμ	Gε	Fγ	Eβ
V	Eε	Fδ	Gκ	Hλ	Aι	Bϑ	Cβ	Dγ	Iα	Jμ	Kζ	Lη
VI	Fζ	Eγ	Hι	Gμ	Bκ	Aη	Dα	Cδ	Jβ	Iλ	Lε	Kϑ
VII	Gη	Hβ	Eμ	Fι	Cλ	Dζ	Aδ	Bα	Kγ	Lκ	Iϑ	Jε
VIII	Hϑ	Gα	Fλ	Eκ	Dμ	Cε	Bγ	Aβ	Lδ	Kι	Jη	Iζ
IX	Iι	Jϑ	Kβ	Lγ	Eα	Fμ	Gζ	Hη	Aε	Bδ	Cκ	Dλ
X	Jκ	Iη	Lα	Kδ	Fβ	Eλ	Hε	Gϑ	Bζ	Aγ	Dι	Cμ
XI	Kλ	Lζ	Iδ	Jα	Gγ	Hκ	Eϑ	Fε	Cη	Dβ	Aμ	Bι
XII	Lμ	Kε	Jγ	Iβ	Hδ	Gι	Fη	Eζ	Dϑ	Cα	Bλ	Aκ

Appendix E
SCT Scenarios

1. EXAMPLES OF SCT SCENARIOS

Szenario 3: Information on flights

your name: *Parker*

intended journey: London Heathrow → Düsseldorf

on June 23th,
morning flight,
direct flight preferred

departure : ______________ h

arrival : ______________ h

flight number : ______________

reservation : 1 seat,
: Economy Class

address : 66 Middle Street, Sheffield
☎ 21 08 33

From which airport is it easier to get into Cologne city centre: Düsseldorf or Cologne/Bonn?

____

Szenario 3: Information on flights

your name: *Heathrow flight information*

flight schedule: London Heathrow → Düsseldorf

		Lufthansa	British Airways	Lufthansa
flight number		LH 2615	BA 381	LH 413
London Heathrow	*dep.*	6:30	6:35	8:20
Brussels	*arr.*		7:35	
Brussels	*dep.*		8:00	
Düsseldorf	*dep.*			
	arr.	7:35	9:05	9:25
		(daily)	*(daily)*	(daily)

reservation:

name : ______________________

address : ______________________

____

telephone number : ______________________

number of seats : ______________________

class ☐ Business ☐ Economy

____

Szenario 11: pizza service

your name: *Clemence*

1 large pizza

for 2 people,
vegetarian pizza preferred

topping : ______________________________

____

price : ______ £

delivery to : 41 Industry Street
Oxford
☎: 7 34 20

How long will it take?

____

Szenario 11: pizza service

your name: *Pizzeria Roma*

pizza

pizzas	1 person	2 persons	4 persons
Toscana (ham, mushrooms, tomatoes, cheese)	£ 3.20	£ 5.95	£ 10.50
Tonno (tuna, onions, tomatoes, cheese)	£ 3.95	£ 7.50	£ 13.95
Fabrizio (salami, ham, tomatoes, cheese)	£ 4.20	£ 7.95	£ 14.95
Vegetaria (spinach, mushrooms, tomatoes, cheese)	£ 4.50	£ 8.50	£ 15.95

delivery to: name : ____________________

address : ____________________

____

telephone number : ____________________

____

2. EXPLANATIONS OF THE SCT DIALOG STRUCTURE GIVEN TO THE TEST SUBJECTS

If you are the caller you have the following symbols:

This symbol means: you are the caller. Please wait until the supervisor tells you to start the conversation.

Next to this symbol the reason for your call is given (eg. I would like to buy a ticket).

*Next to this symbol is a stipulation which should be applied to the exchange of information (eg. I would like to buy a ticket→**but** as cheap as possible).*

Next to this symbol you should write down what information you want to recieve from your partner.

Next to this symbol is all the information your partner requires and which you will pass on to him/her.

If somebody rings you, you have the following symbols:

This symbol means: you are being called. Wait until it rings and pick up the receiver.

Next to this symbol is information from which you should select the details which your partner requires (eg. train ticket prices for adults, children, special conditions for students).

Next to this symbol you should write down all the information you require from your partner.

Both speakers have this symbol

Next to this symbol is a question for which neither you nor your partner will have information. You should discuss the question briefly and find a solution that is acceptable to both of you.

Appendix F
Closing Questionnaire Given after Laboratory Tests

Questions Relating to General Phoning Attitude

- How often do you make a phone call? (very often, often, medium, rarely, very rarely)
- How long do your calls last on average? (very long, long, medium, short, very short)
- Do you like making phone calls? (like very much, like, indifferent, dislike, dislike very much)
- How often do you call foreign countries? (very often, often, medium, rarely, very rarely)

Questions Relating to Cordless Phones

- Do you use a cordless phone?
- If yes, how often do you use it? (very often, often, medium, rarely, very rarely)

Questions Relating to Cellular Phones

- Do you use a cellular phone?
- If yes, how often do you use it? (very often, often, medium, rarely, very rarely)
- If not, how often do you phone other people who have a cellular phone?
- If not, do you plan to buy a cellular phone?
- How would you judge the general quality of a cellular phone in comparison to a wirebound phone? (much better, better, equal, worse, much worse)
- Which advantages or disadvantages does a cellular phone have for you personally?
- Did you experience these advantages/disadvantages in the test?

Questions Relating to an Internet Terminal

- Do you use a computer? (professional, private, studies, none)
- If yes, for how many hours per week?
- Do you like to use a computer? (like very much, like, indifferent, dislike, dislike very much)
- Do you have experience in using the internet? (professional, private, studies, no experience)
- Do you have private internet access?
- Have you ever used an internet phone? (professional, private, studies, never)
- If yes, were there any differences from the system you used today?
- If yes, which overall impression did you have of that system? (excellent, good, fair, poor, bad)
- Would you use an internet phone in your everyday life?
- If yes, for which kind of phone calls do you regard an internet phone appropriate?
- Which expectations do you have of an internet phone?

Questions Relating to the Price

- How much would you pay for a unit of your conventional wirebound phone (cellular phone) in general? (answers in currency per time unit, currency per minute, "points" per minute, etc.)
- Imagine you had the possibility to pay a lower price, but this would also involve having to tolerate a lower quality. Would you take up this offer? Please specify why.

Appendix G
E–Model Algorithm

The main formula in order to calculate the transmission rating factor R is Equation 5.19. The calculation of the additive terms is as follows:

The basic signal–to–noise ration Ro is defined by

$$Ro = 15 - 1.5\,(SLR + No) \tag{G.1}$$

where No [dBm0p] is calculated by power addition of the different noise sources:

$$No = 10 \cdot log_{10}[10^{Nc/10} + 10^{Nos/10} + 10^{Nor/10} + 10^{Nfo/10}] \tag{G.2}$$

Nc [dBm0p] is the sum of all circuit noise powers, referred to the 0 dBr point. Nos [dBm0p] is the equivalent circuit noise at the 0 dBr point, caused by room noise at the send side of level Ps [dB(A)]:

$$Nos = Ps - SLR - Ds - 100 + 0.008\,(Ps - OLR - Ds - 14)^2 \tag{G.3}$$

In the same way, an equivalent circuit noise for the room noise Pr [dB(A)] at the receive side is calculated:

$$Nor = RLR - 121 + Pre + 0.008\,(Pre - 35)^2 \tag{G.4}$$

The term Pre [dBm0p] represents Pr modified by the listener sidetone:

$$Pre = Pr + 10\,log_{10}\,[1 + 10^{(10-LSTR)/10}] \tag{G.5}$$

The noise floor, $Nfor = -64dBmp$, is referred to the 0 dBr point

$$Nfo = Nfor + RLR \tag{G.6}$$

The simultaneous impairment factor, Is, is constituted by three parts:

$$Is = Iolr + Ist + Iq \tag{G.7}$$

$Iolr$ represents the degradation due to an excessively loud connection, given by

$$Iolr = 20\,[\{1 + (X/8)^8\}^{1/8} - X/8] \tag{G.8}$$

with

$$X = OLR + 0.2\,(64 + No - RLR) \qquad \text{(G.9)}$$

Ist covers the degradations due to a non–optimum sidetone:

$$\begin{aligned} Ist = \quad & 10\,[1 + \{(STMRo - 12)/5\}^6]^{1/6} \\ & -46\,[1 + \{STMRo/23\}^{10}]^{1/10} + 36 \end{aligned} \qquad \text{(G.10)}$$

where

$$STMRo = -10\,log_{10}\,[10^{-STMR/10} + e^{-T/4}\,10^{-TELR/10}] \qquad \text{(G.11)}$$

T [ms] and $TELR$ being the one–way delay and loudness rating of talker echo respectively.

The impairment caused by quantizing distortion is expressed by

$$Iq = 15\,log_{10}\,[1 + 10^Y] \qquad \text{(G.12)}$$

where

$$Y = \frac{Ro - 100}{15} + \frac{46 - G}{10} \qquad \text{(G.13)}$$

and G is the equivalent continuous circuit noise, as determined by Richards (1973) to be

$$G = 1.07 + 0.258\,Q + 0.0602\,Q^2 \qquad \text{(G.14)}$$

Q being the signal–to–quantizing–noise ratio, expressed in terms of ‘quantizing distortion units’, qdu (which are used in network planning)

$$Q = 37 - 15\,log_{10}\,(qdu) \qquad \text{(G.15)}$$

This law has been empirically determined and shown to be fulfilled in several tests undertaken in the 1980s (Coleman et al., 1988; South and Usai, 1992). 1 qdu corresponds to the quantizing distortion introduced by one PCM codec.

Also the degradations occurring delayed with respect to the original voice signal are composed of three terms

$$Id = Idte + Idle + Idd \qquad \text{(G.16)}$$

$Idte$ represents the impairments due to talker echo with a one–way delay T [ms] and a loudness rating $TELR$, including masking by noise:

$$Idte = [(Roe - Re)/2 + \sqrt{(Roe - Re)^2/4 + 100} - 1]\,(1 - e^{-T}) \qquad \text{(G.17)}$$

where

$$Roe = -1.5\,(No - RLR) \tag{G.18}$$

$$Re = 80 + 2.5\,(TERV - 14) \tag{G.19}$$

$$TERV = TELR - 40\,log_{10}\,\frac{1 + T/10}{1 + T/150} + 6\,e^{-0.3T^2} \tag{G.20}$$

For $T < 1$ ms, talker echo should be regarded as sidetone, and $Idte = 0$. For very loud sidetone ($STMR < 9$ dB), some talker echo may become masked, and $TERV$ is replaced by

$$TERVs = TERV + Ist/2 \tag{G.21}$$

For very quiet sidetone ($STMR > 15$ dB), the talker echo may become more noticeable, and $Idte$ is replaced by

$$Idtes = \sqrt{Idte^2 + Ist^2} \tag{G.22}$$

The factor $Idle$ takes into account listener echo degradation:

$$Idle = (Ro - Rle)/2 + \sqrt{(Ro - Rle)^2/4 + 169} \tag{G.23}$$

with

$$Rle = 10.5\,(WEPL + 7)(Tr + 1)^{-0.25} \tag{G.24}$$

The impairment due to a long absolute delay is represented by

$$Idd = \begin{cases} 0 & \text{for } Ta \leq 100 \text{ ms} \\ 25\,\{(1 + X^6)^{1/6} - 3\,(1 + [X/3]^6)^{1/6} + 2\} & \text{for } Ta > 100 \text{ ms} \end{cases} \tag{G.25}$$

where

$$X = \frac{log_{10}(Ta/100)}{log_{10}\,2} \tag{G.26}$$

The values for the equipment impairment factor Ie are tabulated for different codecs e.g. in Table 1 of ITU–T Rec. G.107 (1998). Provisional values for the expectation factor A are as in Table 5.2.

The values of R calculated in this way can be transformed into estimations of user opinion in terms of %GoB, %PoW, and MOS, using the Gaussian error function

$$E(X) = \frac{1}{\sqrt{2\pi}} \int_{-\infty}^{x} e^{-\frac{t^2}{2}}\,dt \tag{G.27}$$

and the following equations:

$$\%GoB = 100 \cdot E\left(\frac{R - 60}{16}\right) [\%] \tag{G.28}$$

$$\%PoW = 100 \cdot E\left(\frac{45 - R}{16}\right) [\%] \tag{G.29}$$

$$MOS = \begin{cases} 1 & \text{for } R < 0 \\ 1 + 0.035R + R(R - 60)(100 - R) \cdot 7 \cdot 10^{-6} & \text{for } 0 \leq R \leq 100 \\ 4.5 & \text{for } R > 100 \end{cases} \tag{G.30}$$

Appendix H
Test Conditions and Results

1. TEST CONDITIONS OF THE ISOPREFERENCE TEST

Table H.1. List of test conditions in direct and indirect comparison experiments.

no.	circuit	Ps dB(A)	frequency charact.	indirect comp.				direct comp.	
				LOT	SCT	LOT	SCT	MNRU	TFW
1	MNRU 5	-	handset	X	X	X	X		X
2	MNRU 10	-	handset	X		X			X
3	MNRU 15	-	handset	X		X	X		X
4	MNRU 20	-	handset		X	X	X		X
5	MNRU 25	-	handset	X					X
6	MNRU 35	-	handset	X	X	X	X		X
1a	TFW 0.4	-	handset					X	
2a	TFW 0.3	-	handset					X	
3a	TFW 0.2	-	handset					X	
4a	TFW 0.1	-	handset					X	
5a	TFW 0.05	-	handset					X	
6a	TFW 0.025	-	handset					X	
7	711	-	handset			X	X	X	X
8	711	-	HP -15 dB			X	X	X	X
9	711	-	LP -20 dB			X	X	X	X
10	711	-	BP 0.5-2.5			X	X	X	X
11	711	-	BP 0.8-1.6			X	X	X	X
12	711*726(32)	50	handset	X	X			X	X
13	711*729	50	handset	X	X			X	X
14	726(24)*IS-54	50	handset	X	X			X	X
15	711*726(32)	65	handset	X	X			X	X
16	711*729	65	handset	X	X			X	X
17	726(24)*IS-54	65	handset	X	X			X	X

2. RELATION BETWEEN E–MODEL PREDICTIONS AND TEST RESULTS

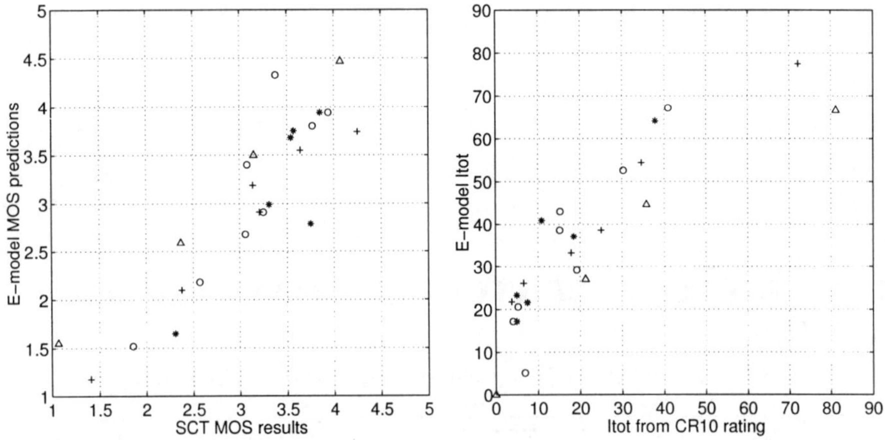

Figure H.1. Comparison of SCT results and E–model predictions, test for short delay talker echo and sidetone. Left: MOS; right: $Itot$. Δ: MNRU; $*$: $STMR = 0$ dB; o: $STMR = 15$ dB; $+$: $STMR = 25$ dB.

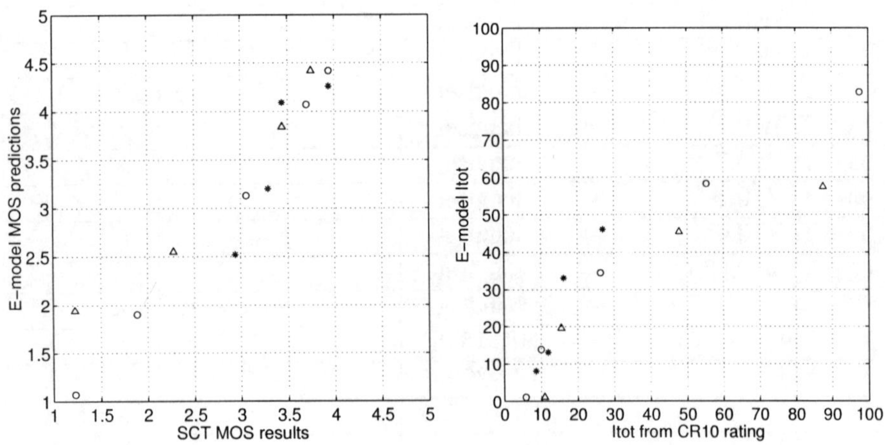

Figure H.2. Comparison of SCT results and E–model predictions, test for talker echo and codecs. Left: MOS; right: $Itot$. Δ: MNRU; $*$: codecs; o: talker echo.

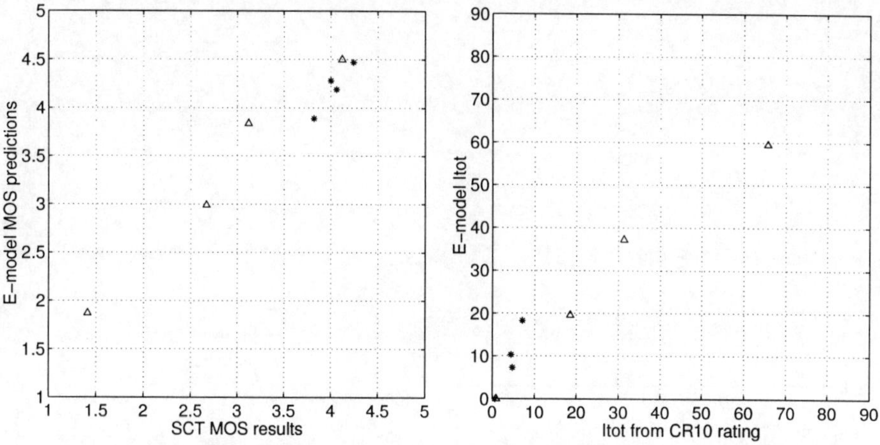

Figure H.3. Comparison of SCT results and E–model predictions, test for short absolute delay ($Ta \leq 200$ ms) and codecs. Left: MOS; right: $Itot$. Δ: MNRU; $*$: codecs.

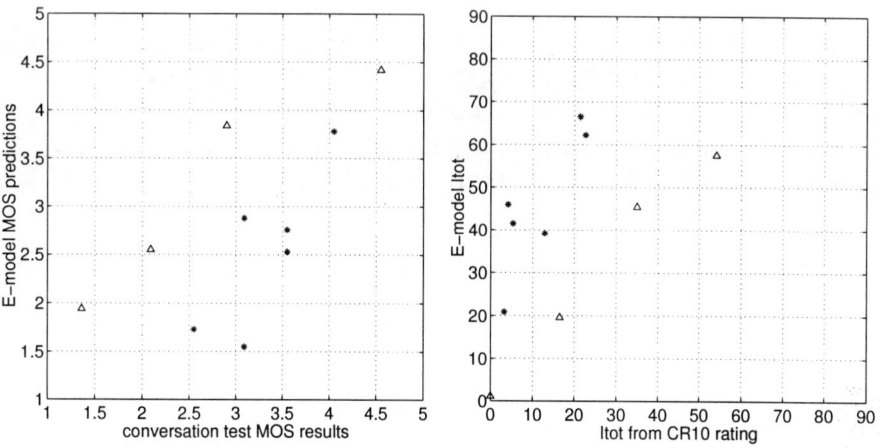

Figure H.4. Comparison of SSRCT results and E–model predictions, test for talker echo and codecs. Left: MOS; right: $Itot$. Δ: MNRU; $*$: talker echo and codecs.

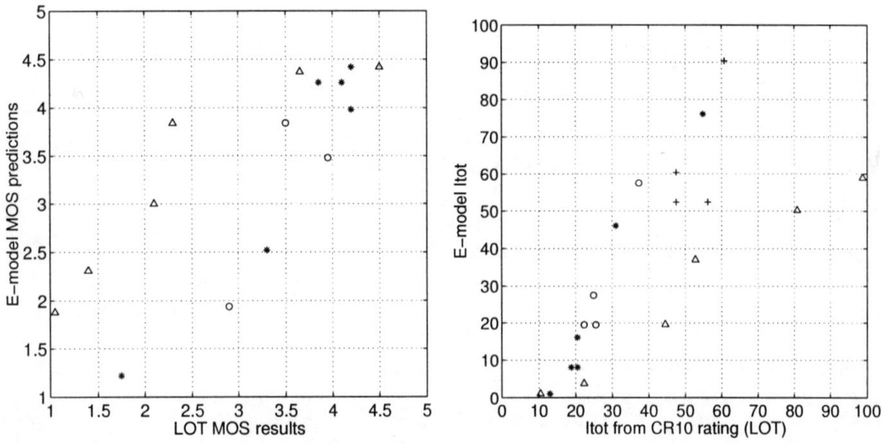

Figure H.5. Comparison of LOT results and E–model predictions, test for codecs and room noise Ps. Left: MOS; right: $Itot$. Δ: MNRU; $*$: codecs; o: $Ps = 50$ dB(A).

3. COMPARISON OF SUBMOD MODEL PREDICTIONS AND TEST RESULTS

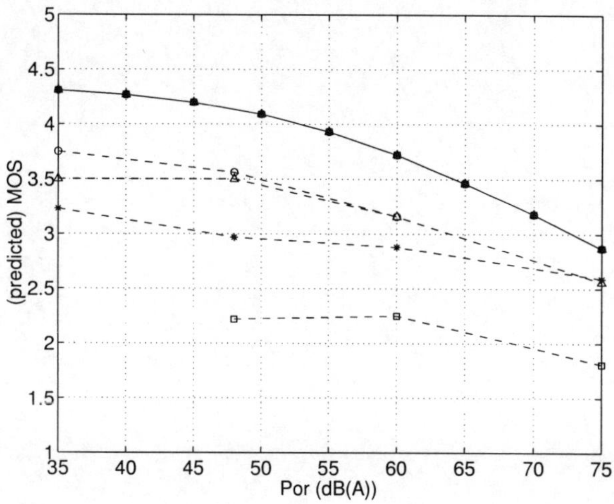

Figure H.6. Comparison of the SUBMOD model (Y_C, solid line) and LOT results (dashed line) for room noise at the send and receive sides (hoth noise). o: Ps = 35 dB(A); Δ: Ps = 48 dB(A); *: Ps = 60 dB(A); □: Ps = 75 dB(A). Test parameters: D = 1, $LSTR$ = 16 dB, $TELR$ = 50 dB, $WEPL$ = 80 dB; all other parameters according to ITU–T Rec. G.107 (1998), Table 3.

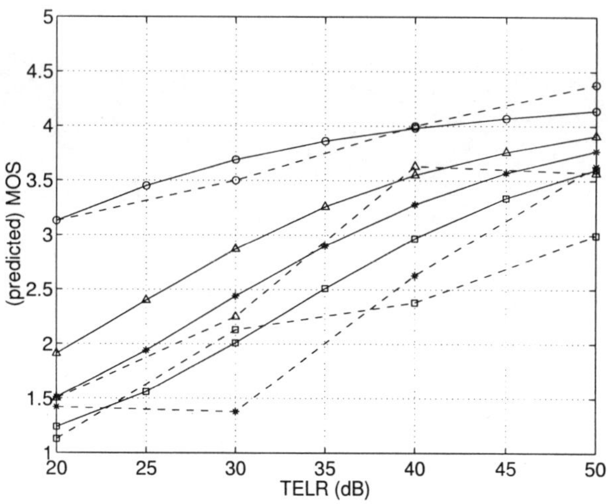

Figure H.7. Comparison of the SUBMOD model (Y_C, solid line) and SCT results (dashed line) for talker echo. o: T = 10 ms; Δ: T = 50 ms; *: T = 100 ms; □: T = 200 ms. Test parameters: D = 1, $LSTR$ = 16 dB, $WEPL$ = 80 dB; all other parameters according to ITU–T Rec. G.107 (1998), Table 3.

Bibliography

Allnatt, J. W. (1975). Subjective Rating and Apparent Magnitude. *Int. J. Man Machine Studies*, 7, 801–816.

Arden, P. and Reeves, B. (1996). Some Considerations for Performing 'Informal' Expert Subjective Assessment. *In: Workshop on Speech Quality Assessment*, BT Laboratories, UK–Martlesham.

Asanuma, N. and Nagabuchi, H. (1994). A New Reference Signal for Evaluating the Quality of Speech Coded at Low Bit Rates. *Electronics and Communications in Japan, Pt. 3*, 77(5), 39–45.

Bappert, V. and Blauert, J. (1994). Auditory Quality Evaluation of Speech–Coding Systems. *acta acustica*, 2, 49–58.

Bednarzyk, M. (1999). *Qualitätsbeurteilung der Geräusche industrieller Produkte: Der Stand der Forschung, abgehandelt am Beispiel der KFZ–Innenraumgeräusche*. Doctoral dissertation, Ruhr–Universität, D–Bochum.

Bennett, R. W. and Lincoln, C. E. (1985). Measuring the Quality of Voice Circuits for Telecommunications Applications. *Proc. 11th Int. Symp. Human Factors in Telephony*, E3, 1–8, F–Cesson Sevigne.

Beranek, L. L. (1971). *Noise and Vibration Control*. McGraw–Hill, USA–New York NY.

Berger, J. (1998). *Instrumentelle Verfahren zur Sprachqualitätsschätzung — Modelle auditiver Tests*. Doctoral dissertation, Christian–Albrechts–Universität Kiel (Arbeiten über Digitale Signalverarbeitung No. 13, U. Heute, ed.), Shaker Verlag, D–Aachen.

Bernsen, N. O., Dybkjaer, H., and Dybkjaer, L. (1996). Principles for the Design of Cooperative Spoken Human–Machine Dialogue. *In: Proc. 4th*

Int. Conf. on Spoken Language Processing ICSLP 96, Vol. 2 (H.T. Bunnell and W. Idsardi, eds.), 729–732, IEEE, USA–Piscataway NJ.

Black, J. W. (1957). Multiple–Choice Intelligibility Tests. *Journal of Speech and Hearing Disorders*, 22, 213–235.

Blauert, J. (1997). *Spatial Hearing: The Psychophysics of Human Sound Localization.* The MIT Press, USA–Cambridge MA.

Bodden, M. and Jekosch, U. (1996). *Entwicklung und Durchführung von Tests mit Versuchspersonen zur Verifizierung von Modellen zur Berechnung der Sprachübertragungsqualität.* Final report to a project funded by Deutsche Telekom AG (unpublished), Institut für Kommunikationsakustik, Ruhr–Universität, D–Bochum.

Borg, G. (1982). A Category Scale with Ratio Properties for Intermodal and Interindividual Comparisons. *In: Psychophysical Judgment and the Process of Perception (H.-G. Geissler and P. Petzold, eds.)*, 25–34, VEB Deutscher Verlag der Wissenschaften, D–Berlin.

Borg, G. (1992). A Fixed Star for Interprocess Comparisons. *In: Fechner Day. Proc. 8th Ann. Meeting of the Int. Soc. for Psychophysics (G. Borg and G. Neely, eds.)*, 41–45, S–Stockholm.

Borg, G. (1994). Psychophysical Scaling: An Overview. *In: Touch, Temperature, and Pain in Health and Disease: Mechanisms and Assessments. Progress in Pain Research and Management, Vol. 3 (J. Boivie, P. Hansson and U. Lindblom, eds.)*, 27–39, IASP Press, USA–Seattle.

Borg, G. (1998). *Borg's Perceived Exertion and Pain Scales.* Human Kinetics, USA–Champaign IL.

Borg, I. and Staufenbiel, T. (1993). *Theorien und Methoden der Skalierung.* Methoden der Psychologie No. 11 (Kurt Pawlik, ed.). Verlag Hans Huber, CH–Bern.

Bortz, J. (1993). *Statistik für Sozialwissenschaftler*, Vol. 4. Springer, D–Berlin.

Bücklein, R. (1962). Hörbarkeit von Unregelmäßigkeiten in Frequenzgängen bei akustischer Übertragung. *Frequenz*, 16/1962, 103–108.

Cavanaugh, J. R., Hatch, R. W., and Sullivan, J. L. (1976). Models for the Subjective Effects of Loss, Noise, and Talker Echo on Telephone Connections. *Bell System Technical Journal*, 55(9), 1319–1371.

Coleman, A. E., Gleiss, N., and Usai, P. (1988). A Subjective Testing Methodology for Evaluating Medium Rate Codecs for Digital Mobile Radio Applications. *Speech Communication*, 7, 151–166.

Collard, J. (1929). A Theoretical Study of the Articulation and Intelligibility of a Telephone Circuit. *Electr. Commun.*, 7, 168–186.

Cotton, B. (1992). New Reference Condition for Very Low Bit Rate Voice Coder Evaluation. *In: Communication for Global Users, IEEE Global Telecommunications Conf. GLOBECOM '92 (USA–Orlando)*, 3, 1719–1722, USA–Piscataway NJ.

Darmoe, P. (1998). *Vergleichende Beurteilung von Sprachübertragungsqualität bezogen auf einen absoluten und relativen Qualitätshorizont.* Study project report (unpublished), Institut für Kommunikationsakustik, Ruhr–Universität, D–Bochum.

Dimolitsas, S. and Ravishankar, C. (1995). Correlation between Headphone and Telephone–Handset Listener Opinion Scores for Single–Stimulus Voice Coder Performance Assessments. *IEEE Signal Processing Letters*, 2(3), 41–43.

DIN 45631 (1991). *Berechnung des Lautstärkepegels und der Lautheit aus dem Geräuschspektrum. Verfahren nach E. Zwicker.* Beuth Verlag, D–Berlin.

Draft ETSI Guide EG/STQ-00004 (1998). *Public Switched Telephone Network: Transmission Delay; Hypothetical Reference Model and Planning Guidelines.* European Telecommunications Standards Institute, F–Sophia Antipolis.

Duncanson, J. P. (1969). The Average Telephone Call Is Better than the Average Telephone Call. *The Public Opinion Quarterly*, 33(1), 112–116.

Duncanson, J. P. (date unknown). *Context Effects in Categorical Judgment.* Unpublished paper.

Dürrer, B. and Jekosch, U. (1998). Meaning of Sounds: A Contribution to Product Sound Design. *In: Proc. 3rd Europ. Conf. On Noise Control EURONOISE 98, Vol. 1 (H. Fastl, ed.)*, 535–540, Deutsche Ges. Akust., D–München.

ETSI Draft Technical Report TIPHON-05001 (1998). *Telecommunications and Internet Protocol Harmonization Over Networks (TIPHON); General Aspects of Quality of Service.* European Telecommunications Standards Institute, F–Sophia Antipolis. Version 1.2.3.

ETSI Technical Report ETR 051 (1992). *Human Factors (HF); Usability Checklist for Telephones. Basic Requirements.* European Telecommunications Standards Institute, F–Sophia Antipolis.

ETSI Technical Report ETR 095 (1993). *Human Factors (HF); Guide for Usability Evaluations of Telecommunication Systems and Services*. European Telecommunications Standards Institute, F–Sophia Antipolis.

ETSI Technical Report ETR 147 (1994). *Human Factors (HF); Usability Checklist for Integrated Services Digital Network (ISDN) Telephone Terminal Equipment*. European Telecommunications Standards Institute, F–Sophia Antipolis.

ETSI Technical Report ETR 250 (1996). *Transmission and Multiplexing (TM); Speech Communication Quality from Mouth to Ear for 3,1 kHz Handset Telephony across Networks*. European Telecommunications Standards Institute, F–Sophia Antipolis.

EURESCOM Project P.807 Deliverable 1 (1998). *Jupiter II - Usability, Performability and Interoperability Trials in Europe*. European Institute for Research and Strategic Studies in Telecommunications, D–Heidelberg.

Fairbanks, C. (1958). Test of Phonemic Differentiation: The Rhyme Test. *J. Acoust. Soc. Am.*, 30(7), 596–600.

Fletcher, H. and Munson, W. (1937). Relation Between Loudness and Masking. *J. Acoust. Soc. Am.*, 9(1), 1–10.

Gabrielsson, A., Schenkman, B. N., and Hagerman, B. (1985). *The Effects of Different Frequency Responses on Sound Quality Judgments and Speech Intelligibility*. Report Technical Audiology No. 112, Karolinska Institutet, Dept. of Technical Audiology, KTH, S–Stockholm.

Gibbon, D. (1992). *EUROM.1 German Speech Database*. ESPRIT project 2589 report (SAM, Multi–Lingual Speech Input/Output Assessment, Methodology and Standardization), Universität Bielefeld, D–Bielefeld.

Gibbon, D., Moore, R., and Winski, R. (1997). *Handbook on Standards and Resources for Spoken Language Systems*. Mouton de Gruyter, D–Berlin.

Gierlich, H. W. (1996). Handys im Akustik–Test. *Funkschau*, 96(20), 22–29.

Gleiss, N. (1970). The Effect of Bandwidth Restriction on Speech Transmission Quality in Telephony. *Proc. 4th Int. Symp. on Human Factors in Telephony (D–Bad Wiessee, 1968)*, 1–6, VDE–Verlag, D–Berlin.

Gleiss, N. (1987a). *Improvements of Rating Methods*. Document SQ-41.87 to ITU-T Speech Quality Experts Group (unpublished), International Telecommunication Union, CH–Geneva.

Gleiss, N. (1987b). *Proposal for Improvements of Supplement 14.* Document SQ-20.87 to ITU-T Speech Quality Experts Group (unpublished), International Telecommunication Union, CH–Geneva.

Gleiss, N. (1989). Desirable Sending Frequency Response of Telephone Sets. *TELE (English edition)*, 1/89, 18–23, Swedish Telecommunications Administration, S–Stockholm.

Gleiss, N. (1992). Usability – Concepts and Evaluation. *TELE (English edition)*, 2/92, 24–30, Swedish Telecommunications Administration, S–Stockholm.

Goodmann, D. J. and Nash, R. D. (1982). Subjective Quality of the Same Speech Transmission Conditions in Seven Different Countries. *In: Proc. ICASSP '82 (F–Paris), Vol. 2*, 984–987, IEEE, USA–New York NY.

Guilford, J. P. (1954). *Psychometric Methods.* McGraw Hill, USA–New York NY.

Halka, U. (1993). *Objektive Qualitätsbeurteilung von Sprachkodiersystemen unter Anwendung von Sprachmodellprozessen.* Doctoral dissertation, Ruhr–Universität Bochum (Arbeiten über Digitale Signalverarbeitung No. 3, U. Heute, ed.), Shaker Verlag, D–Aachen.

Hauenstein, M. (1997). *Psychoakustisch motivierte Maße zur instrumentellen Sprachgütebeurteilung.* Doctoral dissertation, Christian–Albrechts–Universität Kiel (Arbeiten über Digitale Signalverarbeitung No. 10, U. Heute, ed.), Shaker Verlag, D–Aachen.

Hecker, M. H. L. and Williams, C. E. (1966). Choice of Reference Conditions for Speech Preference Tests. *J. Acoust. Soc. Am.*, 39(5), 946–952.

Heinrichs, R. (1993). *Entwicklung von Testverfahren zur Beurteilung charakteristischer akustischer Übertragungseigenschaften von Freisprecheinrichtungen.* Diploma thesis (unpublished), Institut für Kommunikationsakustik, Ruhr–Universität, D–Bochum.

Heller, O. (1990). Scaling and Orientation. *In: Fechner Day. Proc. 6th Ann. Meeting of the Int. Soc. for Psychophysics (F. Müller, ed.)*, 52–57, D–Würzburg.

Hollier, M. P. and Cosier, G. (1996). Assessing Human Perception. *BT Technology Journal*, 14(1), 206–215.

Hoth, D. F. (1941). Room Noise Spectra at Subscribers' Telephone Locations. *J. Acoust. Soc. Am*, 12, 499–504.

House, A. S., Williams, C. E., Hecker, M. H. L., and Kryter, K. D. (1965). Articulation Testing Methods: Consonantal Differentiation with a Closed Response Set. *J. Acoust. Soc. Am.*, 37(1), 158–166.

IEC Publ. 268-13 (date unknown). *Listening Test on Loudspeakers.* International Electrotechnical Commission, CH–Geneva.

IEC Technical Report 60268-13 (1998). *Listening Test on Loudspeakers.* International Electrotechnical Commission, CH–Geneva.

IEEE (1969). IEEE Recommended Practice for Speech Quality Measurements. *IEEE Trans. Audio and Electroacoustics*, AU-17(3), 225–246.

ISO R 532 B (1975). *Acoustics – Method for Calculating Loudness Level.* International Organization for Standardization, CH–Geneva.

ITU–T Contribution COM 12–15 (1997). *Wideband Loudness Rating Algorithm.* Source: MPT China (M. Cao). International Telecommunication Union, CH–Geneva.

ITU–T Contribution COM 12–158 (1983). *Subjective Effects of Sidetone.* Source: American Telephone and Telegraph Company. International Telecommunication Union, CH–Geneva.

ITU–T Contribution COM 12–34 (1997). *TOSQA — Telecommunication Objective Speech Quality Assessment.* Source: Federal Republic of Germany (J. Berger). International Telecommunication Union, CH–Geneva.

ITU–T Contribution COM 12–35 (1997). *Development of Scenarios for a Short Conversation Test.* Source: Federal Republic of Germany (S. Möller). International Telecommunication Union, CH–Geneva.

ITU–T Contribution COM 12–37 (1997). *The E–Model: An Analysis of the Sources and Comparison with Published and New Test Results.* Source: Federal Republic of Germany (S. Möller). International Telecommunication Union, CH–Geneva.

ITU–T Contribution COM 12–5 (1997). *Subjective Evaluation Procedures for Hands–Free Telephones — Double Talk Performance.* Source: Federal Republic of Germany (H. W. Gierlich). International Telecommunication Union, CH–Geneva.

ITU–T Contribution COM 12–62 (1990). *Subjective Evaluation of Pure Delay on Echo–Free Telephone Connections.* Source: Bellcore (USA). International Telecommunication Union, CH–Geneva.

ITU–T Contribution COM 12–65 (1998). *Subjective Evaluation of Echo Cancellers with Short Conversation Test.* Source: France Telecom/CNET (D. Pascal). International Telecommunication Union, CH–Geneva.

ITU–T Contribution COM 12–69 (1998). *E–Model Predictions and the Impairment Factor Principle for Low–Bitrate Codecs and Quantizing Distortion: Analysis of Test Results.* Source: Federal Republic of Germany (S. Möller). International Telecommunication Union, CH–Geneva.

ITU–T Contribution COM 12–70 (1998). *E–Model Predictions for Room Noise at Send Side: Proposal for Modeling the Lombard Effect.* Source: Federal Republic of Germany (S. Möller). International Telecommunication Union, CH–Geneva.

ITU–T Contribution COM 12–76 (1996). *Experiences with the ETSI Model Planning Evolving Private Networks.* Source: TELENORMA GmbH/Bosch Telecom (J. Pomy). International Telecommunication Union, CH–Geneva.

ITU–T Contribution COM 12–79 (1996). *Influence of Room Noise at the Send Side: An Analysis of Data Taken from Supplement 32 (Series G).* Source: Ericsson (N. O. Johannesson). International Telecommunication Union, CH–Geneva.

ITU–T Contribution COM 12–94 (1991). *Conversational Test on GSM DMR System + Olympus Satellite. Effect of Delay, Errors and Echo.* Source: CSELT (Italy). International Telecommunication Union, CH–Geneva.

ITU–T Delayed Contribution D.009 (1997). *User Perception of Total Quality for Wired vs. Wireless Telephones.* Source: CANADA/NORTEL (L. Thorpe). International Telecommunication Union, CH–Geneva.

ITU–T Delayed Contribution D.014 (1997). *Digital–Signal–Processing–Based Procedure for Impulsive Noise Detection.* Source: NECSY/ITALTEL (P. Paglierani). International Telecommunication Union, CH–Geneva.

ITU–T Delayed Contribution D.023 (1997). *Comparison of Results of Different Subjective Evaluation Procedures for Hands–Free Terminals.* Source: Deutsche Telekom AG (H. W. Gierlich). International Telecommunication Union, CH–Geneva.

ITU–T Delayed Contribution D.069 (1998). *Accuracy Analysis of Mapping Nonintrusive Measurements to the E–Model.* Source: AT&T (D. B. Ramsden). International Telecommunication Union, CH–Geneva.

ITU–T Delayed Contribution D.088 (1998). *Draft Text for Proposed Recommendation P.56X – Analysis and Interpretation of INMD Voice–Service Mea-*

surements. Source: British Telecom (P. Sheppard). International Telecommunication Union, CH–Geneva.

ITU–T Delayed Contribution D.093 (1998). *Application of E–Model to North American Transmission Planning Scenarios.* Source: Nortel Networks/Canada (P. Coverdale). International Telecommunication Union, CH–Geneva.

ITU–T Draft Rec. G.109 (1998). *Definition of Categories of Speech Transmission Quality.* International Telecommunication Union, CH–Geneva.

ITU–T Draft Rec. P.831 (1998). *Subjective Performance Evaluation of Network Echo Cancellers.* International Telecommunication Union, CH–Geneva.

ITU–T Handbook on Telephonometry (1992). International Telecommunication Union, CH–Geneva.

ITU–T Rec. E.125 (1998). *Inquiries among Users of the International Telephone Service (Red Book edition).* International Telecommunication Union, CH–Geneva.

ITU–T Rec. E.800 (1994). *Terms and Definitions Related to Quality of Service and Network Performance Including Dependability.* International Telecommunication Union, CH–Geneva.

ITU–T Rec. G.107 (1998). *The E–Model, a Computational Model for Use in Transmission Planning.* International Telecommunication Union, CH–Geneva.

ITU–T Rec. G.113 (1996). *Transmission Impairments.* International Telecommunication Union, CH–Geneva.

ITU–T Rec. G.114 (1996). *One–Way Transmission Time.* International Telecommunication Union, CH–Geneva.

ITU–T Rec. G.175 (1997). *Transmission Planning for Private/Public Network Interconnection of Voice Traffic.* International Telecommunication Union, CH–Geneva.

ITU–T Rec. G.712 (1992). *Transmission Performance Characteristics of Pulse Code Modulation.* International Telecommunication Union, CH–Geneva.

ITU–T Rec. H.323 (1998). *Packet–Based Multimedia Communication Systems.* International Telecommunication Union, CH–Geneva.

ITU–T Rec. O.41 (1994). *Psophometer for Use on Telephone–Type Circuits.* International Telecommunication Union, CH–Geneva.

ITU–T Rec. P.11 (1993). *Effect of Transmission Impairments.* International Telecommunication Union, CH–Geneva.

ITU–T Rec. P.48 (1989). *Specifications for an Intermediate Reference System (Blue Book edition).* International Telecommunication Union, CH–Geneva.

ITU–T Rec. P.64 (1997). *Determination of Sensitivity/Frequency Characteristics of Local Telephone Systems.* International Telecommunication Union, CH–Geneva.

ITU–T Rec. P.76 (1989). *Determination of Loudness Ratings; Fundamental Principles (Blue Book edition).* International Telecommunication Union, CH–Geneva.

ITU–T Rec. P.78 (1996). *Subjective Testing Method for Determination of Loudness Ratings in Accordance with Recommendation P.76.* International Telecommunication Union, CH–Geneva.

ITU–T Rec. P.79 (1993). *Calculation of Loudness Ratings for Telephone Sets.* International Telecommunication Union, CH–Geneva.

ITU–T Rec. P.800 (1996). *Methods for Subjective Determination of Transmission Quality.* International Telecommunication Union, CH–Geneva.

ITU–T Rec. P.810 (1996). *Modulated Noise Reference Unit (MNRU).* International Telecommunication Union, CH–Geneva.

ITU–T Rec. P.82 (1989). *Method for Evaluation of Service from the Standpoint of Speech Transmission Quality (Blue Book edition).* International Telecommunication Union, CH–Geneva.

ITU–T Rec. P.861 (1996). *Objective Quality Measurement of Telephone–Band (300–3400 Hz) Speech Codecs.* International Telecommunication Union, CH–Geneva.

ITU–T Suppl. 3 to P–Series Rec. (1993). *Models for Predicting Transmission Quality from Objective Measurements.* International Telecommunication Union, CH–Geneva.

ITU–T Suppl. 5 to P–Series Rec. (1985). *The SYBIL Method of Subjective Testing (Red Book edition).* International Telecommunication Union, CH–Geneva. (No longer in force).

Jekosch, U. (1992). The Cluster–Identification Test. *In: Proc. 2th Int. Conf. on Spoken Language Processing (ICSLP'92)*, 1, 205–209, CDN–Banff.

Jekosch, U. (1998). *Sprache hören und beurteilen. Qualitätsbeurteilung von Sprechtechnologien als Forschungs– und Dienstleistungsaufgabe*. Habilitation thesis, Universität/Gesamthochschule, D–Essen.

Johannesson, N. O. (1996). *Echo Canceller Performance Characterized by Impairment Factors*. Document IP–16 to ITU–T Speech Quality Experts Group (unpublished), International Telecommunication Union, (presented at the UK–Ipswich meeting, Sept. 23–27), CH–Geneva.

Johannesson, N. O. (1997). The ETSI Computation Model: A Tool for Transmission Planning of Telephone Networks. *IEEE Communications Magazine*, Jan., 70–79.

Johnston, R. D. (1996). Beyond Intelligibility — The Performance of Text–To–Speech Synthesisers. *BT Technology Journal*, 14(1), 100–111.

Johnston, R. D. (1997). Towards a Reference System for Speech Technology. *In: Proc. Speech and Language Technology (SALT) Club Workshop on Evaluation in Speech and Language Technology*, June 17–18, 90–97, UK–Sheffield.

Jones, B. L. and McManus, P. R. (1986). Graphic Scaling of Qualitative Terms. *J. Soc. of Motion Picture and Television Eng.*, 95, 1166–1171.

Karis, D. (1991). Evaluating Transmission Quality in Mobile Telecommunication Systems Using Conversation Tests. *Proc. of the Human Factors Soc. 35th Ann. Meeting*, 217–221, USA–San Francisco CA.

Kettler, F., Hottenbacher, A., Heinrichs, R., and Gierlich, H. W. (1994). Verfahren zur subjektiven Qualitätsbeurteilung von Freisprecheinrichtungen. *In: Fortschritte der Akustik – DAGA '94*, 1365–1368, DPG GmbH, D–Bad Honnef.

Kitawaki, N. and Itoh, K. (1990). Delay Effect Assessment Taking into Account of Human Factors in Telecommunications. *Proc. 13th Int. Symp. on Human Factors in Telecommunications*, 555–562, I–Turin.

Kitawaki, N. and Itoh, K. (1991). Pure Delay Effects on Speech Quality in Telecommunications. *IEEE J. on Selected Areas in Communications*, 9(4), 586–593.

Kliche, I., Kapust, R., and Noll, P. (1997). Ein offener modularer Simulator für Untersuchungen zur Sprachübertragung über ATM. *In: Elektronische Sprachsignalverarbeitung*, Studientexte zur Sprachkommunikation No. 14, 184–191, Gesellschaft für Signalverarbeitung und Mustererkennung, D–Dresden.

Kort, B. W. (1983). Models and Methods for Evaluating Customer Acceptance of Telephone Connections. *In: World Communication Year – A Time For Planning, IEEE Conf. GLOBECOM '83*, 706–714, USA–Piscataway NJ.

Krebber, W. (1995). *Sprachübertragungsqualität von Fernsprech–Handapparaten*, Vol. 357. Series 10, VDI–Verlag GmbH, D–Düsseldorf.

Lane, H. L., Catania, A. C., and Stevens, S. S. (1961). Voice Level: Autophonic Scale, Perceived Loudness, and Effects of Sidetone. *J. Acoust. Soc. Am.*, 33(2), 160–167.

Lane, H. L., Tranel, B., and Sisson, C. (1970). Regulation of Voice Communication by Sensory Dynamics. *J. Acoust. Soc. Am.*, 47(2), 618–624.

Lewandowski, T. (1980). *Linguistisches Wörterbuch*, Vol. 3. Quelle und Meyer, D–Heidelberg.

Lewis, N. W. and Allnatt, J. W. (1965). Subjective Quality of Television Pictures with Multiple Impairments. *Electronics Letters*, 1(7), 187–188.

McDermott, B. J. (1969). Multidimensional Analyses of Circuit Quality Judgments. *J. Acoust. Soc. Am.*, 45(3), 774–781.

McGee, V. E. (1964). Semantic Components of the Quality of Processed Speech. *J. of Speech and Hearing Research*, 7, 310–323.

McGee, V. E. (1965). Determining Perceptual Spaces for the Quality of Filtered Speech. *J. of Speech and Hearing Research*, 8, 23–38.

McKelvie, S. J. (1978). Graphic Rating Scales — How Many Categories? *Br. J. Psychol.*, 69, 185–202.

Möller, S. and Berger, J. (1998). Vergleichende Untersuchung von Prädiktionsmodellen zur Sprachübertragungsqualität. *Sprachkommunikation, ITG–Fachbericht 152 (R. Hoffmann, ed.)*, 89–92, VDE–Verlag GmbH, D–Berlin.

Möller, S. and Jekosch, U. (1998). *Verifikation des ETSI–Modells zur Bestimmung der Sprachübertragungsqualität im Zusammenhang unterschiedlicher Systemkonfigurationen*. Final report to a project funded by Deutsche Telekom Berkom GmbH (unpublished), Institut für Kommunikationsakustik, Ruhr–Universität, D–Bochum.

Müller, F., Schmidt, A., and Paulsen, R. (1996). The Influence of Stimulus Range on the Judgments of Loudness and Annoyance. *Proc. Internoise '96*, 2095–2098, Institute of Acoustics, UK–St. Albans.

Munson, W. A. and Karlin, J. E. (1962). Isopreference Method for Evaluating Speech–Transmission Circuits. *J. Acoust. Soc. Am.*, 34(6), 762–774.

Nakatani, L. H. and Dukes, K. D. (1973). A Sensitive Test of Speech Communication Quality. *J. Acoust. Soc. Am.*, 53(4), 1083–1092.

Ogden, C. K. and Richards, I. A. (1960). *The Meaning of Meaning (10th edition).* Routledge & Kegan Paul Ltd., UK–London.

Osgood, C. E., Suci, G., and Tannenbaum, P. (1957). *The Measurement of Meaning.* University of Illinois Press, USA–Urbana.

Panzer, I. L. and Sharpley, A. D. (1991). Comparison of Three Methods for the Subjective Measurement of Speech Quality. *J. Acoust., Soc. Am.*, 90(4), 2327.

Parducci, A. (1963). Range–Frequency Compromise in Judgment. *Psychological Monographs*, 77(2), 1–50.

Pascal, D. and Boyer, M. (1990). Multidimensional Perceptive Measurement of Quality: Comparative Performance of Two Methods. *Proc. 13th Int. Symp. on Human Factors in Telecommunications*, 519–520, I–Turin.

Peirce, C. S. (1998). *Phänomen und Logik der Zeichen (H. Pape, Ed.).* Suhrkamp, D–Frankfurt/M.

Plücker, R. (1998). *Untersuchung des Einflusses der Qualitätserwartung und der Telefonier–Situation auf die Beurteilung von Telefonsprache.* Study project report (unpublished), Institut für Kommunikationsakustik, Ruhr–Universität, D–Bochum.

Quackenbush, S. R., Barnwell, T. P., and Clemens, M. A. (1988). *Objective Measures of Speech Quality.* Prentice Hall, USA–Englewood Cliffs NJ.

Raghavarao, D. (1971). *Constructions and Combinatorial Problems in Design of Experiments.* John Wiley & Sons, USA–New York NY.

Richards, D. L. (1973). *Telecommunication by Speech.* Butterworths, UK–London.

Richards, D. L. (1974). Calculation of Opinion Scores for Telephone Connections. *Proc. IEE*, 121(5), 313–323.

Riedel, J. (1998). *Einfluß der Erwartungshaltung des Benutzers auf die Qualitätsbeurteilung von Telefonübertragungen an Internet–Terminals.* Study project report (unpublished), Institut für Kommunikationsakustik, Ruhr–Universität, D–Bochum.

Rosenberger, J. R. (1988). Quality Assessment Methods for Speech Coding. *Telecommunication Journal*, 55, 820–825.

Rothauser, E. H., Urbanek, G. E., and Pachl, W. P. (1968). Isopreference Method for Speech Evaluation. *J. Acoust. Soc. Am.*, 44(2), 408–418.

Schäfer, B. (1983). *Semantische Differential Technik*, Enzyklopädie der Psychologie, 154–221. Verlag für Psychologie Dr. C. J. Hogrefe, D–Göttingen.

Schulte-Fortkamp, B. (1994). *Geräusche beurteilen im Labor*, Vol. 100. Series 17, VDI–Verlag GmbH, D–Düsseldorf.

Sixtl, F. (1967). *Meßmethoden der Psychologie*. Julius Beltz, D–Weinheim.

Sotscheck, J. (1982). Ein Reimtest für Verständlichkeitsmessungen mit deutscher Sprache als ein verbessertes Verfahren zur Bestimmung der Sprachübertragungsgüte. *Der Fernmeldeingenieur*, 36(4/5), 1–84.

South, C. and Usai, P. (1992). Subjective Performance of CCITT's 16 kbit/s LD–CELP Algorithm with Voice Signals. *In: Communication for Global Users, IEEE Global Telecommunications Conf. GLOBECOM '92 (USA–Orlando)*, 1709–1714, USA–Piscataway NJ.

Stevens, S. S. (1946). On the Theory of Scales of Measurement. *Science*, 103(2684), 677–680.

Stevens, S. S. (1957). On the Psychophysical Law. *Psychol. Rev.*, 64, 153–181.

Stevens, S. S. and Galanter, E. H. (1957). Ratio Scales and Category Scales for a Dozen Perceptual Continua. *J. Exp. Psychol.*, 54, 377–411.

Thomas, I. (1968). The Influence of First and Second Formants on the Intelligibility of Clipped Speech. *J. Audio Eng. Soc.*, 16, 182–185.

Thurstone, L. L. (1927). A Law of Comparative Judgment. *Psychological Review*, 34, 273–286.

Verschoor, A. (1994). Quality of Speech Services in GSM – Definition of a Universal Quality Parameter for Quality Comparison and Evaluation. *In: Wireless Networks Catching the Mobile Future. Proc. 5th Int. Symp. on Personal, Indoor and Mobile Radio Communications (PIMRC '94)*, Vol. 4, 1348–1352, IOS Press, NL–Amsterdam.

Voiers, W. D. (1977). Diagnostic Acceptability Measure for Speech Communication Systems. *In: Proc. ICASSP '77 (USA–Hartford)*, 204–207, IEEE, USA–New York NY.

Voiers, W. D., Sharpley, A., and Hehmsoth, C. (1975). *Research on Diagnostic Evaluation of Speech Intelligibility*. Research Repüort AFCRL-72-0694, Air Force Cambridge Research Laboratories, USA–Bedford MA.

Wiegelmann, S. (1997). *Einsatzmöglichkeiten und –grenzen der MNRU als Referenzverzerrung bei der Bestimmung von Sprachübertragungsqualität im Telefoniebereich.* Diploma thesis (unpublished), Institut für Kommunikationsakustik, Ruhr–Universität, D–Bochum.

Wilkie, W. L. (1994). *Consumer Behaviour.* John Wiley & Sons Inc., USA–New York NY.

Williams, G. and Barnes, G. J. (1974). The Specification and Measurement of Sidetone as Dictated by Human Factor Considerations. *Proc. 7th Int. Symp. on Human Factors in Telecommunications (R. W. Archbold, ed.)*, CND–Montreal.

Zwicker, E. (1961). Subdivision of the Audible Frequency Range into Critical Bands (Frequenzgruppen). *J. Acoust. Soc. Am.*, 33(2), 248.

Zwicker, E. and Fastl, H. (1999). *Psychoacoustics: Facts and Models.* Springer, D–Berlin.

Zwislocki, J. J. and Goodman, D. A. (1980). Absolute Scaling of Sensory Magnitudes: A Validation. *Perception & Psychophysics*, 28(1), 28–38.

Index